Lecture Notes in Mathematics 1688

Springer-Verlag Berlin Heidelberg GmbH

Sigurd Assing Wolfgang M. Schmidt

Continuous Strong Markov Processes in Dimension One

A Stochastic Calculus Approach

Springer

Authors

Sigurd Assing
Universität Bielefeld
Fakultät für Mathematik
Postfach 100131
D-33501 Bielefeld

Wolfgang M. Schmidt
Deutsche Bank AG
OTC-Derivate
Große Gallusstraße 10–14
D-60272 Frankfurt

Cataloging-in-Publication Data applied for

Die Deutsche Bibliothek - CIP-Einheitsaufnahme

Assing, Sigurd:
Continuous strong Markov processes in dimension one : a stochastic calculus approach / Sigurd Assing ; Wolfgang Schmidt. - Berlin ; Heidelberg ; New York ; Barcelona ; Budapest ; Hong Kong ; London ; Milan ; Paris ; Santa Clara ; Singapore ; Tokyo : Springer, 1998
(Lecture notes in mathematics ; 1688)
ISBN 3-540-64465-2

Mathematics Subject Classification (1991): 60J25, 60G44, 60H20

ISSN 0075-8434
ISBN 978-3-540-64465-1 ISBN 978-3-540-69786-2 (eBook)
DOI 10.1007/978-3-540-69786-2

Typesetting: Camera-ready $\TeX$ output by the authors
SPIN: 10649856 46/3143-543210 - Printed on acid-free paper

To our sons
Edgar, Gregor and Sebastian

Preface

The purpose of this book is the investigation of one-dimensional continuous strong Markov processes using methods of stochastic calculus.

Markov processes are characterized by the principle of independence of future and past given the present value of the process, which goes back to A.A. MARKOV. The class of Markov processes plays an outstanding role in applications of random processes since many time dependent random phenomena can suitably be modelled by them. On the other hand, due to its manifold connections to other fields of mathematics, as for instance functional analysis, potential theory and partial differential equations, the theory of Markov processes is one of the most interesting branches of the theory of random processes.

There are two basic approaches to the investigation of Markov processes.

The analytical approach, which goes back in substantial part to A.N. KOLMOGOROV, is based on the investigation of partial differential equations for the transition densities. Here a Markov process is characterized by its infinitesimal generator of the semigroup of linear operators generated by the transition probabilities. Particularly, W. FELLER [25], [26], [27], E.B. DYNKIN [16] and K. ITÔ, H.P. MCKEAN [31] substantially contributed to this approach.

The second approach is based on the construction of paths of Markov processes out of paths of well-understood more elementary processes such as the Wiener and the Poisson process. In his pioneering article [30] in 1951 K. ITÔ constructed Markov processes as solutions of stochastic differential equations. To this purpose he generalized the stochastic integral of N. WIENER for the Wiener process to random integrands.

The work of K. ITÔ can be seen as one of the roots of modern stochastic calculus. In the sequel the notion of the stochastic integral as a central object of stochastic analysis was extended to square integrable martingales by H. KUNITA, S. WATANABE [36] and finally to semimartingales by the Strasbourg school. A semimartingale is a random process admitting a decomposition into the sum of a local martingale and a process with paths of finite variation. The development of the (one-dimensional, adapted) integral culminated in the fundamental result due to K. BICHTELER [4],[5] and C. DELLACHERIE [13] that every "reasonable" stochastic integrator is necessarily a semimartingale. At the same time, the class of semimartingales is very large, it contains for example processes with independent increments and "many" Markov processes.

In the present work we restrict our attention to one-dimensional continuous (homogeneous) strong Markov processes. Such processes are often called diffusions.

For regular diffusions W. FELLER [25], [26], [27] described the explicit structure of the infinitesimal generator. The monograph of K. ITÔ, H.P. MCKEAN [31] contains an exhausting study of infinitesimal generators even for singular diffusions. V.A. VOLKONSKI [46], E.B. DYNKIN [16] and K. ITÔ, H.P. MCKEAN [31] gave a constructive description of paths of regular diffusions. Starting from a Wiener process they constructed regular diffusions by random time change, spatial transformation and killing. Using methods of stochastic calculus S. MÉLÉARD [38] investigated regular strong Markov processes with killing. Also L.C.G. ROGERS, D. WILLIAMS [41] contains a detailed study of regular diffusions in the framework of stochastic calculus.

It should be emphasized that up to now regular diffusions have been studied most widely and almost solely. Roughly speaking, a diffusion is called regular if in a neighbourhood of each inner point of its state interval it behaves like a Wiener process up to a continuous time change and a space transformation.

One purpose of this work is to remove the restriction of regularity and to give a unified probabilistic approach to a systematic treatment of *arbitrary* one-dimensional continuous strong Markov processes.

The study of the relations between Markov processes and martingales or semimartingales is up to now a constant source of interest, and there are various papers devoted to this subject, see, e. g., J.L. DOOB [15], E. CINLAR, J. JACOD, P. PROTTER, M.J. SHARPE [11], E. CINLAR, J. JACOD [10], H.J. ENGELBERT, W. SCHMIDT [21], [22], [23], [24]. In their paper [10], CINLAR and JACOD proved that every semimartingale Hunt process is a nice time change of a so called Itô process. Thereby an Itô process is a solution of a stochastic differential equation in the sense of K. ITÔ [30]. However the above time change is in no way unique and, therefore, from this one can hardly derive results on the structure of the underlying process. So the problem arises to describe the semimartingale decomposition of a continuous strong Markov semimartingale itself and to give conditions under which it turns out to be a solution of a certain stochastic differential equation. Among other topics, these problems will be solved in this book. For the case of regular semimartingale diffusions some of the problems were treated in [43]. However, as already pointed out, one of the main interests of this book is to eliminate the condition of regularity completely.

Now let us have a more detailed look at the contents of the book and its organisation.

Chapter I briefly reviews basic definitions and results from the theory of Markov processes and from stochastic calculus.

Chapter II provides the well-known classification of points of the state interval of a continuous strong Markov process into regular, right and left singular points. We refine this classification and give a suitable decomposition of the state interval. On this basis some basic results on the behaviour of a continuous strong Markov process in its singular and regular points are proved.

In Chapter III we introduce the notion of a weakly additive functional. This notion is of fundamental importance for what follows. Roughly speaking, a right continuous increasing process is called weakly additive if it is, in contrast to a perfect additive functional, only additive in its points of right-increase. Changing time in a strong Markov process by the right-inverse of a certain weakly additive functional inherits the strong Markov property. For the wide class of right processes this result was already published in [44].

As indicated above, the method of random time change plays an outstanding role in the study of regular diffusions. There, the time changes are right-inverses of certain perfect additive functionals. Introducing the notion of a weakly additive functional, we are able to depart from the restriction of regularity of the continuous strong Markov process under consideration.

In Chapter IV we investigate the decomposition of a continuous strong Markov semimartingale into its local martingale part and its part of finite variation. The explicit form of the latter can be calculated explicitly, it is the sum of an integral and sums over the local time of the underlying process and two deterministic functions of the occupation time of the process in certain subsets of the singular points.

Chapter V introduces the speed measure of an arbitrary continuous strong Markov process. We derive an extremely useful formula for the occupation time of the process in measurable subsets of the state interval. As it is known, the speed measure plays an important role in the investigation of regular diffusions. Here we extend the definition of the speed measure in an appropriate way to arbitrary diffusions. Moreover, in Chapter V we show that under a slight technical condition every continuous strong Markov process can be transformed into a continuous strong Markov semimartingale by a one-to-one transformation of the state interval.

Chapter VI is devoted to the construction of continuous strong Markov processes. Our intention is to give a general approach to the construction of arbitrary continuous strong Markov processes. Thereby the notion of a weakly additive functional introduced in Chapter III plays a central role. The results of Chapter VI generalize the well-known construction of regular diffusions due to VOLKONSKI [46], ITÔ, MCKEAN [31] in that we do not restrict ourselves to regular diffusions.

In Chapter VII we consider the problem under which conditions a given continuous strong Markov process turns out to be a solution of a certain stochastic differential equation. We give simple necessary and sufficient conditions on the speed measure and the scale function guaranteeing that the given process is a solution of a stochastic differential equation with generalized drift or a stochastic differential equation in the sense of K. ITÔ. As a consequence we show that the class of Itô processes already contains a large number of non-regular diffusions whose drift and diffusion functions are, however, highly irregular. Several examples illuminate fundamental cases of non-regular diffusions.

There are two Appendices providing some auxiliary results on the behaviour of functionals of the Wiener process and certain signed measures that are also of independent interest.

We are grateful to H.J. ENGELBERT (Jena) who taught us stochastic processes and stochastic calculus. Several results of this book are further developments of ideas that originate from earlier joint work with him.

Table of Contents

I. Basic Concepts and Preparatory Results

In this chapter we introduce some important notions as well as specific results which are basic to our approach. For all unexplained notions, further details and related topics, the reader is referred to such books as [6], [8], [14], [16], [32], [35], [40], [41], [49].

I.1 Basic Probability Structure

Let I denote an interval of the real line $\mathbb{R}$, $\mathfrak{B}(I)$ the σ-algebra of Borel subsets of I and $\mathfrak{B}^u(I)$ its universal completion.

(1.1) Throughout this book we consider a measurable space $(\Omega, \mathcal{F}^0)$ equipped with a family $(\mathbf{P}_x, x \in I)$ of probability measures such that for every $A \in \mathcal{F}^0$ the function $\mathbf{P}_{.}(A)$ is $\mathfrak{B}^u(I)$-measurable. Then, for every probability measure μ on $(I, \mathfrak{B}^u(I))$, we define a probability measure $\mathbf{P}_\mu$ on $\mathcal{F}^0$ by setting

$$\mathbf{P}_\mu(A) = \int_I \mathbf{P}_x(A)\, \mu(\mathrm{d}x), \qquad A \in \mathcal{F}^0.$$

Let $\mathcal{F}^\mu$ be the completion of $\mathcal{F}^0$ under $\mathbf{P}_\mu$. The collection of all $\mathbf{P}_\mu$-null sets generates a σ-algebra which is denoted by $\mathcal{N}(\mathbf{P}_\mu)$. We set $\mathcal{F} = \bigcap \mathcal{F}^\mu$ where the intersection is over all probability measures μ on $(I, \mathfrak{B}(I))$. The measures $\mathbf{P}_\mu$, and, in particular, the measures $\mathbf{P}_x$, can be extended to $\mathcal{F}$ in an obvious way. This leads to the family $(\Omega, \mathcal{F}, \mathbf{P}_x, x \in I)$ of probability spaces. Finally, we say that an assertion holds almost surely (a.s.) if it holds $\mathbf{P}_x$-a.s. for all $x \in I$.

(1.2) On the family $(\Omega, \mathcal{F}, \mathbf{P}_x, x \in I)$ introduced in (1.1) we consider a real-valued stochastic process $X = (X)_{t \geq 0}$ equipped with a filtration $\mathbb{F} = (\mathcal{F}_t)_{t \geq 0}$ of sub-σ-algebras of $\mathcal{F}$, and satisfying the following assumptions:

(i) $$\mathbf{P}_x(\{X_0 = x\}) = 1, \qquad x \in I.$$

(ii) X is $\mathbb{F}$-adapted, which is indicated by the notation $(X, \mathbb{F})$,

(iii) $\mathcal{F}_0$ is augmented in $\mathcal{F}$ with respect to the family $(\mathbf{P}_\mu)$ (see [6], I(5.3)).

(1.3) Definition The stochastic process X is said to possess *shift operators* if there exists a semigroup $\Theta = (\theta_t)_{0 \le t \le \infty}$ of operators $\theta_t : \Omega \to \Omega$ such that

$$X_{t+s} = X_t \circ \theta_s, \qquad 0 \le t < \infty,\ s \ge 0, \text{ a.s.}$$

and

$$X_t \circ \theta_\infty = X_\infty \quad \text{on} \quad \{X_\infty = \lim_{t \to \infty} X_t \text{ exists and is finite}\}, \qquad t \ge 0, \text{ a.s.}$$

Let the σ-algebra $\mathcal{F}_t^0$ be given by $\mathcal{F}_t^0 = \sigma\{X_s : s \le t\}$ and write $\mathcal{F}_t^X$ for the augmentation of $\mathcal{F}_t^0$ in $\mathcal{F}$ with respect to $(\mathbf{P}_\mu)$. The filtration $(\mathcal{F}_t^X)_{t \ge 0}$ is denoted by $\mathbb{F}^X$.

For a filtration $\mathbb{H} = (\mathcal{H}_t)_{t \ge 0}$, we denote by $\mathbb{H}_+$ the smallest right-continuous filtration containing $\mathbb{H}$. We write $\mathcal{H}_\infty$ for the σ-algebra $\bigvee_{t \ge 0} \mathcal{H}_t$. For two filtrations $\mathbb{H} = (\mathcal{H}_t)_{t \ge 0}$ and $\mathbb{G} = (\mathcal{G}_t)_{t \ge 0}$ we denote $\mathbb{H} \vee \mathbb{G} = (\mathcal{H}_t \vee \mathcal{G}_t)_{t \ge 0}$.

Let $B \in \mathfrak{B}(I)$. We define the first entry time of X in B by

$$\mathrm{D}_B(X) = \inf\{s \ge 0 : X_s \in B\},$$

where, by convention, $\inf \emptyset$ is set to infinity. In what follows, all processes $(X, \mathbb{F})$ under consideration are progressively measurable, and, therefore, $\mathrm{D}_B(X)$ is an $\mathbb{F}_+$-stopping time. If there is no danger of confusion, we frequently use the notation D_B instead of $\mathrm{D}_B(X)$.

Finally, given a measurable process X and an $\mathbb{F}$-stopping time T the stopped process $(X_{t \wedge T})_{t \ge 0}$ is denoted by X^T.

I.2 Strong Markov Processes

Let $(X, \mathbb{F})$ on $(\Omega, \mathcal{F}, \mathbf{P}_x, x \in I)$ be a continuous stochastic process as in (1.2) and taking values in I.

(1.4) Definition $(X, \mathbb{F})$ is said to be a *continuous strong Markov process with state space I* if the following conditions are satisfied:

(i) X possesses shift operators $\Theta = (\theta_t)_{t \ge 0}$.
(ii) For every nonnegative $\mathcal{F}_\infty^X$-measurable random variable Y it holds that

$$\mathbf{E}_x(Y \circ \theta_t | \mathcal{F}_t) = \mathbf{E}_{X_t}(Y) \qquad \mathbf{P}_x - \text{a.s.}$$

(iii) For every $\mathbb{F}_+$-stopping time T and every nonnegative $\mathcal{F}_\infty^X$-measurable random variable Y it holds that [1]

$$\mathbf{E}_x(Y \circ \theta_T \mathbf{1}_{\{T < \infty\}} | \mathcal{F}_{T+}^X) = \mathbf{E}_{X_T}(Y) \mathbf{1}_{\{T < \infty\}} \qquad \mathbf{P}_x - \text{a.s.}$$

[1] $\mathbf{1}_A$ denotes the indicator function of a set A.

As is well-known, for (ii) and (iii) it suffices to take $Y = \mathbf{1}_{\{X_s \in B\}}$, $s \geq 0$, $B \in \mathfrak{B}(I)$.

(1.5) Remark If $(X, \mathbb{F})$ is a strong Markov process on $(\Omega, \mathcal{F}, \mathbf{P}_x, x \in I)$, then $(X, \mathbb{F}_+^X)$ is a strong Markov process as well. This implies (see [6], I(8.12))

$$\mathbb{F}^X = \mathbb{F}_+^X,$$

i.e., $\mathbb{F}^X$ is a right-continuous filtration.

I.3 Semimartingales

Let $(\Omega, \mathcal{F}, \mathbf{P}_x, x \in I)$ be a family of probability spaces as in (1.1) equipped with a filtration $\mathbb{F} = (\mathcal{F}_t)_{t \geq 0}$ of sub-σ-algebras of $\mathcal{F}$. We use the following notation:

$$\begin{aligned}
\mathcal{S}(\mathbb{F}) &= \{Y : (Y, \mathbb{F}) \text{ is a continuous semimartingale with respect to } \mathbf{P}_x \text{ for every } x \in I\},\\
\mathcal{M}(\mathbb{F}) &= \{M : M_0 = 0 \text{ and } (M, \mathbb{F}) \text{ is a continuous local martingale with respect to } \mathbf{P}_x \text{ for every } x \in I\},\\
\mathcal{V}^+(\mathbb{F}) &= \{V : V_0 = 0 \text{ and each path of } (V, \mathbb{F}) \text{ is increasing}\},\\
\mathcal{V}(\mathbb{F}) &= \mathcal{V}^+(\mathbb{F}) - \mathcal{V}^+(\mathbb{F}).
\end{aligned}$$

If T is an $\mathbb{F}$-stopping time, then we further define

$$\mathcal{M}(\mathbb{F}, T) = \{(M, \mathbb{F}) \text{ is a continuous local martingale up to } T, \text{ i.e., there exists a sequence of stopping times } T_n \uparrow T \text{ such that } (M^{T_n}, \mathbb{F}) \text{ is a continuous } \mathbf{P}_x\text{-martingale for all } x \in I\}.$$

By definition, for $Y \in \mathcal{S}(\mathbb{F})$ it follows that for each fixed $x \in I$ there is a decomposition $Y = Y_0 + M^x + V^x$ $\mathbf{P}_x$-a.s. where $(M^x, \mathbb{F})$ is a continuous local martingale and $V^x \in \mathcal{V}(\mathbb{F})$. A priori, the semimartingale decomposition of Y depends on the underlying measure $\mathbf{P}_x$. However, CINLAR, JACOD, PROTTER and SHARPE [11] have shown that under weak assumptions there exists a universal decomposition.

(1.6) Proposition ([11], Th. (3.12); [45]) *If (1.2) is satisfied then for every $Y \in \mathcal{S}(\mathbb{F}_+)$ we have:*

(i) *There exists an $M \in \mathcal{M}(\mathbb{F}_+)$ and a $V \in \mathcal{V}(\mathbb{F}_+)$ such that*

$$Y = Y_0 + M + V \qquad a.s.$$

(ii) *There exists a continuous process $\langle Y \rangle \in \mathcal{V}^+(\mathbb{F}_+)$ which is a version of the quadratic variation of $(M, \mathbb{F}_+)$ with respect to $\mathbf{P}_x$, for all $x \in I$.*

(iii) *Let H be an $\mathbb{F}_+$-predictable process such that the stochastic integral with respect to Y is well-defined for all $x \in I$. Then there exists an $H * Y \in \mathcal{S}(\mathbb{F}_+)$ which is a version of the $\mathbf{P}_x$-stochastic integral $\int_0^{\cdot} H_s \, \mathrm{d}Y_s$ for all $x \in I$.*

Now assume (1.2) and let X possess shift operators $\Theta = (\theta_t)_{t \geq 0}$.

(1.7) Definition (i) A stochastic process Y on $(\Omega, \mathcal{F}, \mathbf{P}_x, x \in I)$ is called *additive* (resp. *perfect additive*) provided $Y'_{t+s} = Y'_s + Y'_t \circ \theta_s$ a.s. for all $s, t \geq 0$ (resp. for all $s, t \geq 0$ a.s.) where $Y' = Y - Y_0$.
(ii) A map $T : \Omega \to [0, \infty]$ is called a *perfect terminal time* provided T is $\mathcal{F}^X_\infty$-measurable and $T \circ \theta_t = T - t$ on $\{T > t\}$, $t \geq 0$, a.s.

The semimartingales we consider in the following are often additive, which allows us to strengthen (1.6).

(1.8) Proposition ([11], Th. (3.18),(3.21); [21]) *Let $Y \in \mathcal{S}(\mathbb{F}^X)$ be additive where $(X, \mathbb{F})$ is a strong Markov process. Then the process M in (1.6) can be chosen to be additive. Furthermore, $\langle Y \rangle$ as well as $V = Y - M$ can be chosen perfectly additive.*

Remark. In (1.8) it is sufficient to assume that $(X, \mathbb{F})$ is a Markov process. However, in the sequel we only investigate strong Markov processes.

(1.9) Definition Let $(X, \mathbb{F})$ on $(\Omega, \mathcal{F}, \mathbf{P}_x, x \in I)$ be a continuous strong Markov process. If $X \in \mathcal{S}(\mathbb{F}^X)$, then we call $(X, \mathbb{F})$ a *continuous strong Markov semimartingale.* The semimartingale decomposition of X according to (1.6) is denoted by

(1.10) $$X = X_0 + M(X) + V(X)$$

where $M(X) \in \mathcal{M}(\mathbb{F}^X)$ and $V(X) \in \mathcal{V}(\mathbb{F}^X)$.

For a continuous strong Markov semimartingale, recall that $\mathbb{F}^X = \mathbb{F}^X_+$ (see (1.5)).

(1.11) Definition A process $(X, \mathbb{F})$ as in (1.2) is called *Wiener process* provided $X - X_0 \in \mathcal{M}(\mathbb{F}_+)$ and $\langle X \rangle_t = t$, $t \geq 0$, a.s.

If a Wiener process $(X, \mathbb{F})$ possesses shift operators, it is well known that it is a strong Markov process in the sense of (1.4), in other words $(X, \mathbb{F})$ is a continuous strong Markov semimartingale (see for example [35], 2.6.15).

I.4 Local Times and the Generalized Itô-formula

Local times of semimartingales play an outstanding role in the study of the "fine" structure of semimartingales because they provide "valuable information" on the behaviour of the trajectories. For semimartingales, the notion of local time was introduced by MEYER in [39]. The reader can find an in-depth study of local times in [3].

We assume (1.1) and (1.2) and let $Y \in \mathcal{S}(\mathbb{F}_+)$ be a semimartingale. We introduce the notion of the local time via the Tanaka-formula, which is one of the most convenient ways. Let $y \in I$ be fixed. For the present, the right local time $\widetilde{L}_+^Y(\cdot, y)$ of Y at the point y is defined by

$$\widetilde{L}_+^Y(\cdot, y) = | Y - y | - | Y_0 - y | - \operatorname{sign}_-(Y) * Y$$

where $\operatorname{sign}_- = -\mathbf{1}_{(-\infty,0]} + \mathbf{1}_{(0,\infty)}$ and the last summand on the right-hand side corresponds with the stochastic integral mentioned in (1.6)(iii). Replacing sign_- by $\operatorname{sign}_+ = -\mathbf{1}_{(-\infty,0)} + \mathbf{1}_{[0,\infty)}$, analogously one can introduce the left local time $\widetilde{L}_-^Y(\cdot, y)$. Clearly, for all $x \in I$ we have that $\widetilde{L}_+^Y(\cdot, y)$ and $\widetilde{L}_-^Y(\cdot, y)$ are versions of the right and left local time of the $\mathbf{P}_x$-semimartingale $(Y, \mathbb{F}_+)$, respectively. Therefore $\widetilde{L}_+^Y(\cdot, y)$ and $\widetilde{L}_-^Y(\cdot, y)$ are $\mathbb{F}_+^Y$-adapted (see also [32], (9.19)). For the local times of a $\mathbf{P}_x$-semimartingale $(Y, \mathbb{F}_+)$ the existence of regular versions is well-known (cf. [51]). The next lemma shows that this remains true for $Y \in \mathcal{S}(\mathbb{F}_+)$.

(1.12) Lemma *Let $Y \in \mathcal{S}(\mathbb{F}_+)$. There exist $L_+^Y(t, y)$ and $L_-^Y(t, y)$, $(t, y) \in [0, \infty) \times \mathbb{R}$ such that*

(i) *$L_+^Y(\cdot, y)$ and $L_-^Y(\cdot, y)$ are continuous for every $y \in \mathbb{R}$.*

(ii) *It holds that*

$$L_+^Y(t, y) = \lim_{\substack{s \to t \\ x \downarrow y}} L_+^Y(s, x) \quad \textit{and} \quad L_-^Y(t, y) = \lim_{\substack{s \to t \\ x \uparrow y}} L_-^Y(s, x),$$

and there exist the limits

$$\lim_{\substack{s \to t \\ x \uparrow y}} L_+^Y(s, x) \quad \textit{and} \quad \lim_{\substack{s \to t \\ x \downarrow y}} L_-^Y(s, x).$$

(iii) $$L_+^Y(t, y) = \widetilde{L}_+^Y(t, y),\ L_-^Y(t, y) = \widetilde{L}_-^Y(t, y) \quad \text{a.s., } (t, y) \in [0, \infty) \times \mathbb{R}.$$

For the proof of this lemma we refer to [43], Appendix.

In what follows, we always work with the good versions $L_+^Y(\cdot, y)$ and $L_-^Y(\cdot, y)$ of the local times as stated in (1.12). They are unique up to indistinguishability and are called *right* and *left local time of* $Y \in \mathcal{S}(\mathbb{F}_+)$ *at the point* $y \in \mathbb{R}$, respectively. Finally, we define the *(symmetric) local time of* $Y \in \mathcal{S}(\mathbb{F}_+)$ *at the point* $y \in \mathbb{R}$ by

$$\text{(1.13)} \qquad L^Y(\cdot,y) = \frac{1}{2}\left(L_+^Y(\cdot,y) + L_-^Y(\cdot,y)\right).$$

Obviously, all local times of Y at the point $y \in \mathbb{R}$ are $\mathbb{F}_+^Y$-adapted (see also (1.17)(v)).

For convenience, in what follows we make use of the agreement

$$L_+^Y(\cdot,\pm\infty) = L_-^Y(\cdot,\pm\infty) = L^Y(\cdot,\pm\infty) = 0.$$

The following generalization of the "classical" Itô-formula turns out to be very useful.

(1.14) Proposition (Generalized Itô-formula) *Let* $Y \in \mathcal{S}(\mathbb{F}_+)$ *and* $f : \mathbb{R} \to \mathbb{R}$ *be a function that can be represented as the difference of two convex functions. Then* $f(Y) \in \mathcal{S}(\mathbb{F}_+)$ *holds and we have*

$$f(Y) = f(Y_0) + f'(Y) * Y + \frac{1}{2}\int_{\mathbb{R}} L^Y(\cdot,y)\,\mathrm{d}f_+'(y) \qquad a.s.$$

where $f' = \frac{1}{2}(f_+' + f_-')$. *Here* f_+' *and* f_+' *denote the right- and left-hand derivative of* f, *respectively. An analogous formula is true if we replace* f' *by* f_+' *(resp.* f_-'*) and the symmetric local time is replaced by* $L_-^Y(\cdot,y)$ *(resp.* $L_+^Y(\cdot,y)$*).*

For the proof, we refer to [39],[32].

Consider $Y \in \mathcal{S}(\mathbb{F}_+)$ with values in an interval $I \subseteq \mathbb{R}$. A function $f : I \to \mathbb{R}$ is called a *semimartingale function of* Y provided $f(Y) \in \mathcal{S}(\mathbb{F}_+)$. The above proposition implies that every map $f : \mathbb{R} \to \mathbb{R}$ that can be represented as the difference of two convex functions is a semimartingale function for all $Y \in \mathcal{S}(\mathbb{F}_+)$.

Comparing the "classical" and the generalized Itô-formula we get the equality

$$\begin{aligned}\text{(1.15)} \qquad \int_0^t g(Y_s)\,\mathrm{d}\langle Y\rangle_s &= \int_{\mathbb{R}} g(y) L^Y(t,y)\,\mathrm{d}y \\ &= \int_{\mathbb{R}} g(y) L_+^Y(t,y)\,\mathrm{d}y \\ &= \int_{\mathbb{R}} g(y) L_-^Y(t,y)\,\mathrm{d}y,\ t \geq 0, \text{ a.s.,}\end{aligned}$$

for every integrable or nonnegative measurable function g.
In the case of a continuous strong Markov semimartingale $(X,\mathbb{F})$, from (1.15) we can derive the additivity of the local times of X.

(1.16) Lemma (i) *For a continuous strong Markov semimartingale* $(X, \mathbb{F})$ *it holds that*

$$L_\pm^X(t+s, y) = L_\pm^X(t, y) \circ \theta_s + L_\pm^X(s, y), \qquad t, s \geq 0,\ y \in \mathbb{R},\ a.s.$$

In particular, $L^X(\cdot, y)$, $L_+^X(\cdot, y)$, $L_-^X(\cdot, y)$ *are perfect additive.*

(ii) *Further, for every perfect terminal time* T*, we have*

$$L_\pm^X((t+s) \wedge T, y) = L_\pm^X(t \wedge T, y) \circ \theta_s + L_\pm^X(s \wedge T, y), \quad t \geq 0,\ y \in \mathbb{R},$$

on $\{T > s\}$ *for all* $s \geq 0$ *a.s.*

Proof. (i) It suffices to consider the right local time. Using (1.8) from (1.15) we obtain

$$\begin{aligned}
\int_{\mathbb{R}} g(y) L_+^X(t, y)\, \mathrm{d}y \circ \theta_s &= \left(\int_0^t g(X_u)\, \mathrm{d}\langle X \rangle_u \right) \circ \theta_s \\
&= \int_0^t g(X_{u+s})\, \mathrm{d}(\langle X \rangle_{u+s} - \langle X \rangle_s) \\
&= \int_s^{s+t} g(X_u)\, \mathrm{d}\langle X \rangle_u \\
&= \int_{\mathbb{R}} g(y)(L_+^X(t+s, y) - L_+^X(s, y))\, \mathrm{d}y
\end{aligned}$$

for all $t \geq 0$ a.s., every bounded measurable function g and each $s \geq 0$. This implies $L_+^X(t+s, y) = L_+^X(t, y) \circ \theta_s + L_+^X(s, y)$, $t \geq 0$, for Lebesgue-almost all $y \in \mathbb{R}$ a.s. which leads to $L_+^X(t+s, y) = L_+^X(t, y) \circ \theta_s + L_+^X(s, y)$, $y \in \mathbb{R}$, $t \geq 0$, a.s. because $L_+^X(t, \cdot)$ is right-continuous. So we have proved the additivity of $L_+^X(\cdot, y)$ at least for every $y \in \mathbb{R}$. WALSH has shown in [47] that $L_+^X(\cdot, y)$ is indistinguishable from a perfect additive functional $\hat{L}_+^X(\cdot, y)$ where the exceptional set can depend on y. But[2] $\bar{L}_+^X(t, y) = \lim_{\substack{x \in \mathbb{Q} \\ x \downarrow y}} \hat{L}_+^X(t, x)$, $t \geq 0$, $y \in \mathbb{R}$, is indistinguishable from $L_+^X(t, y)$ and satisfies the asserted properties.

(ii) follows directly from (i) and (1.7)(ii). □

The following result collects further properties of the local time.

(1.17) Proposition *For* $Y \in \mathcal{S}(\mathbb{F}_+)$ *with the semimartingale decomposition* $Y = Y_0 + M + V$ *(cf. 1.6(i)) it holds that*

(i) $L^Y(t, y) = 0$ *on* $\{y \notin [\min_{s \leq t} Y_s, \max_{s \leq t} Y_s]\}$, $\quad t \geq 0$, *a.s.*

(ii) *For* $B \in \mathfrak{B}(\mathbb{R})$ *we have* $L^Y(\cdot \wedge D_B(Y), z) = 0$, $z \in B$, *a.s.*

(iii) $\int_0^t \mathbf{1}_{\{y\}}(Y_s) L_\pm^Y(\mathrm{d}s, y) = L_\pm^Y(t, y)$, $\quad t \geq 0$, *a.s.*, *for all* $y \in \mathbb{R}$.

[2] $\mathbb{Q}$ (resp. $\mathbb{Q}_+$) denotes the set of rational (resp. nonnegative rational) numbers.

(iv) *Fix* $y \in \mathbb{R}$ *and assume* $Y_t \geq y$ *(resp.* $Y_t \leq y$*) for all* $t \geq \mathrm{D}_{\{y\}}(Y)$ *a.s. Then* $L_-^Y(\cdot, y) = 0$ *(resp.* $L_+^Y(\cdot, y) = 0$*) a.s.*

(v) *For a fixed* $y \in \mathbb{R}$ *and* $\varepsilon > 0$ *we define*

$$U_0^\varepsilon(Y) = 0,$$

$$V_n^\varepsilon(Y) = \inf\{t \geq U_n^\varepsilon(Y) : Y_t \leq y\},$$

$$U_n^\varepsilon(y) = \inf\{t \geq V_n^\varepsilon(Y) : Y_t \geq y + \varepsilon\},$$

$n = 1, 2, \ldots$ *and denote by*

$$C_t^\varepsilon(Y) = \sum_{n \geq 1} \mathbf{1}_{\{U_n^\varepsilon(Y) \leq t\}}$$

the number of crossings of the interval $[y, y+\varepsilon]$ *by* Y *up to the time* t. *Then for all* $T \geq 0$ *and* $x \in I$ *we have*

$$\sup_{t \leq T} |\varepsilon C_t^\varepsilon(Y) - \frac{1}{2} L_+^Y(t, y)| \to 0$$

in $\mathbf{P}_x$*-probability as* ε *tends to zero. An analogous assertion holds true for the left local time.*

(vi) *Let* $x \in I$ *be fixed,* $\mathbf{P}_x(\{Y_0 = y_0\}) = 1$ *and denote by* $V = V^+ - V^-$ *a suitable decomposition of* $V_\cdot(\omega)$ *into a difference of increasing functions* $V_\cdot^+(\omega)$, $V_\cdot^-(\omega)$ *with disjoint support. Assume that* $\mathbf{P}_x(\{\langle M \rangle_t > 0\}) = 1$, $t > 0$, *and there exists an* $\epsilon > 0$ *such that* $\int_0^t \mathbf{1}_{(y_0 - \epsilon, y_0]}(Y_s) \mathrm{d}V_s^- = 0$, $\int_0^t \mathbf{1}_{(y_0, y_0+\epsilon)}(Y_s) \mathrm{d}V_s^+ = 0$, $t > 0$, $\mathbf{P}_x - a.s.$ *Then*

$$\mathbf{P}_x(\{L_+^Y(t, y_0) > 0\}) = 1, \qquad t > 0.$$

An analogous result holds for $L_-^Y(t, y_0)$ *if we replace* $(y_0 - \epsilon, y_0]$ *by* $(y_0 - \epsilon, y_0)$ *and* $(y_0, y_0 + \epsilon)$ *by* $[y_0, y_0 + \epsilon)$.

Proof. The properties (i),(ii),(iii) are well-known. They can be derived from the generalized Itô-formula (1.14) and (1.15). Property (v) has been proved by ElKaroui in [17], Proposition 2. Item (iv) follows from (v).

It remains to prove (vi). By the generalized Itô-formula

$$| Y_t - y_0 | = | Y_0 - y_0 | + \int_0^t \mathrm{sign}_-(Y_s - y_0) \mathrm{d}M_s + \int_0^t \mathrm{sign}_-(Y_s - y_0)\, \mathrm{d}V_s + L_+^Y(t, y_0).$$

Since $(| Y_\cdot - y_0 |, L_+^Y(\cdot, y_0))$ is the solution of the so-called reflection problem associated with $| Y_0 - y_0 | + \int_0^\cdot \mathrm{sign}_-(Y_s - y_0) \mathrm{d}M_s + \int_0^\cdot \mathrm{sign}_-(Y_s - y_0)\, \mathrm{d}V_s$ (cf. [8], chapter 8), applying a well-known lemma on the representation of the unique solution of a reflection problem (cf. e.g. [7], [8] Lemma 8.1, [35] Lemma 3.6.14) we obtain that $\mathbf{P}_y$-a.s.

$$L_+^Y(t,y_0) = -\min_{0\le u\le t}\left[\int_0^u \mathrm{sign}_-(Y_s-y_0)\mathrm{d}M_s + \int_0^u \mathrm{sign}_-(Y_s-y_0)\mathrm{d}V_s\right].$$

Now

$$\int_0^t \mathrm{sign}_-(Y_s-y_0)\mathrm{d}V_s \le \int_0^t \mathbf{1}_{(-\infty,y_0]}(Y_s)\mathrm{d}V_s^- + \int_0^t \mathbf{1}_{(y_0,\infty)}(Y_s)\mathrm{d}V_s^+.$$

Then, using our assumption for $0<t\le \mathrm{D}_{(y_0-\epsilon,y_0+\epsilon)}(Y)$ we get

$$\begin{aligned} L_+^Y(t,y_0) &\ge -\min_{0\le u\le t}\Big[\int_0^u \mathrm{sign}_-(Y_s-y_0)\mathrm{d}M_s \\ &\quad + \int_0^u \mathbf{1}_{(-\infty,y_0]}(Y_s)\mathrm{d}V_s^- + \int_0^u \mathbf{1}_{(y_0,\infty)}(Y_s)\mathrm{d}V_s^+\Big] \\ &\ge -\min_{0\le u\le t}\left[\int_0^u \mathrm{sign}_-(Y_s-y_0)\mathrm{d}M_s\right] \end{aligned}$$

$\mathbf{P}_x$-a.s. The local martingale $N_u = \int_0^u \mathrm{sign}(Y_s-y_0)\mathrm{d}M_s$ admits the same quadratic variation as M, i.e. $\langle N\rangle = \langle M\rangle$ a.s. Now, according to our assumption, $\mathbf{P}_x(\{\langle N\rangle_t > 0\}) = 1$, $t>0$, and thus $\mathbf{P}_x(\{\min_{0\le u\le t} N_u < 0\}) = 1$, $t>0$, which yields the assertion. □

The next lemma provides a formula to compute the local times if the underlying semimartingale is subject to a space transformation.

(1.18) Lemma *Let $Y\in\mathcal{S}(\mathbb{F}_+)$ be a semimartingale taking values in an interval J whose open kernel is denoted by J^0. Further, let $q: J\to\mathbb{R}$ be an absolutely continuous strictly increasing function with $q(x) = q(x_0) + \int_{x_0}^x q'(y)\,dy$, $x_0\in J$ fixed, where q' is supposed to possess a right-hand limit $q'_+(y)$ and a left-hand limit $q'_-(y)$ at each point $y\in J^0$. Suppose $Z=q(Y)$ is again a semimartingale, i.e. $Z\in\mathcal{S}(\mathbb{F}_+)$, and suppose that*

$$\langle Z\rangle_t = \int_0^t q'_\pm(Y_s)^2\mathbf{1}_{J^0}(Y_s)\,\mathrm{d}\langle Y\rangle_s, \qquad t\ge 0,\ a.s.,$$

is satisfied. Then it holds that

$$L_\pm^Z(\cdot,q(y)) = L_\pm^Y(\cdot,y)q'_\pm(y), \qquad y\in J^0,\ a.s.$$

Moreover, if $L_+^Y(t,\cdot)$ and $L_-^Y(t,\cdot)$ are a.s. continuous at a point $y\in J^0$ we obtain for the symmetric local time the formula

$$L^Z(\cdot,q(y)) = L^Y(\cdot,y)q'(y) \qquad a.s.$$

with $q' = \frac{1}{2}(q'_+ + q'_-)$.

(1.19) Remark If the function $q: J\to\mathbb{R}$ is strictly increasing and a difference of two convex functions then the assumptions of the above lemma immediately follow from the generalized Itô-formula (1.14).

Proof of (1.18). For each nonnegative measurable function g, from (1.15) it follows that

$$\begin{aligned}\int_{\mathbb{R}} g(y)L^Z_{\pm}(t,y)\,\mathrm{d}y &= \int_0^t g(Z_s)\,\mathrm{d}\langle Z\rangle_s \\ &= \int_0^t g(q(Y_s))(q'_{\pm}(Y_s))^2 \mathbf{1}_{J^0}(Y_s)\,\mathrm{d}\langle Y\rangle_s \\ &= \int_{J^0} g(q(y))(q'_{\pm}(y))^2 L^Y_{\pm}(t,y)\,\mathrm{d}y \\ &= \int_{q(J^0)} g(y)\,q'_{\pm}(p(y))L^Y_{\pm}(t,p(y))\,\mathrm{d}y,\end{aligned}$$

$t \geq 0$, a.s., where p denotes the inverse function to q. Hence, we have $L^Z_{\pm}(t,y) = L^Y_{\pm}(t,p(y))\,q'_{\pm}(p(y))$ for Lebesgue-almost all $y \in q(J^0)$, $t \geq 0$, a.s. Using the right and left continuity of both sides, respectively, this equality extends to all $y \in q(J^0)$, $t \geq 0$, a.s., proving the first assertion. If the local times $L^Y_{\pm}(t,\cdot)$ are a.s. continuous at a point $y \in J^0$ then $L^Y_+(t,y) = L^Y_-(t,y) = L^Y(t,y)$ is satisfied (see [51]). Therefore, the second assertion is an easy consequence of the first one and (1.13). □

I.5 Random Time Change

The method of random time change plays an outstanding role in the investigation of paths of one-dimensional stochastic processes and so we shall often utilize this method in what follows.

We start with a family of probability spaces $(\Omega, \mathcal{F}, \mathbf{P}_x, x \in I)$ as in (1.1) and with a filtration $\mathbb{F}$ satisfying (1.2)(iii).

(1.20) Let A be a right-continuous increasing process taking values in $[0,\infty]$ whose *right-inverse process* T is defined by $T_t = \inf\{s \geq 0 : A_s > t\}$, $t \geq 0$. If A is $\mathbb{F}$-adapted, then every T_t is an $\mathbb{F}_+$-stopping time. A right-continuous increasing family of $\mathbb{F}$-stopping times is called an $\mathbb{F}$-*time change*. We set $A_{0-} = 0$ (resp. $T_{0-} = 0$) and say that A (resp. T) is *continuous at the point* 0 provided $A_0 = A_{0-}$ (resp. $T_0 = T_{0-}$).

In order to change the time of a process Y it is convenient to define its value at the "time" infinity:

$$Y_\infty = \begin{cases} \lim_{t\to\infty} Y_t & : \text{ if the limit exists} \\ 0 & : \text{ otherwise.} \end{cases}$$

Now for a measurable process Y and an $\mathbb{F}$-time change T define the *time-changed process* $Y \circ T$ by $(Y_{T_t})_{t\geq 0}$ and the *time-changed filtration* $\mathbb{F} \circ T$ by $(\mathbb{F}_{T_t})_{t\geq 0}$. If Y is $\mathbb{F}$-progressively measurable, it is well-known that $Y \circ T$ is $\mathbb{F} \circ T$-adapted. Finally, let us mention the following result.

(1.21) Lemma *Let $Y \in \mathcal{S}(\mathbb{F}_+)$ and let T be an $\mathbb{F}_+$-time change whose right-inverse is denoted by A. If $A_\infty = \infty$ a.s. then $Y \circ T$ is a $\mathbf{P}_x$-semimartingale for every $x \in I$.*

Proof. Proposition (10.10) in [32] states that $(Y \circ T, \mathbb{F}_+ \circ T)$ is a semimartingale on $\bigcup_n [0, A_n] = [0, \infty)$. □

I.6 Equality of Distributions

The following proposition turns out to be useful for proving that the distribution of two processes coincide. In [20],[22] this method has already been applied to show the uniqueness in law of solutions of stochastic differential equations. We present a slightly more general result.

(1.22) Let C denote the space of continuous functions $\omega : [0, \infty) \to \mathbb{R}$, $\mathcal{C}_t^0 = \sigma\{\omega(s) : s \leq t\}$ and let $\mathcal{C}_\infty^0 = \bigvee_t \mathcal{C}_t^0$. For a family of probability spaces $(C, \mathcal{C}_\infty^0, \mathrm{Q}_x, x \in I)$ as in (1.1), the completion of $\mathcal{C}_\infty^0$ with respect to (Q_μ) is denoted by $\mathcal{C}_\infty(\mathrm{Q})$. We write $\mathbb{C}(\mathrm{Q}) = (\mathcal{C}_t(\mathrm{Q}))_{t \geq 0}$ for the filtration $(\mathcal{C}_t^0)_{t \geq 0}$ augmented in $\mathcal{C}_\infty(\mathrm{Q})$ with respect to (Q_μ) (see [6], I(5.3)).

(1.23) Proposition *Let $(\Omega^i, \mathcal{F}^i, \mathbf{P}_x^i, x \in I)$, $i = 1, 2$, be two families of probability spaces as in (1.1) equipped with filtrations $\mathbb{F}^i$, $i = 1, 2$, satisfying (1.2)(iii). Let X^i be an $\mathbb{F}^i$-adapted process with the following properties:*

(i) *There exist a filtration $\mathbb{G}^i$ of sub-σ-algebras of $\mathcal{F}^i$, a process $(Y^i, \mathbb{G}^i)$ and a $\mathbb{G}_+^i$-time change T^i such that $X^i = Y^i \circ T^i$ a.s., $i = 1, 2$.*

(ii) *Y^1 and Y^2 possess the same distribution, i.e., the image measures $\mathbf{P}_x^1 \circ (Y^1)^{-1}$ and $\mathbf{P}_x^2 \circ (Y^2)^{-1}$ coincide on $\mathcal{C}_\infty^0$ for each $x \in I$: $\mathbf{P}_x^1 \circ (Y^1)^{-1} = \mathbf{P}_x^2 \circ (Y^2)^{-1} =: \mathrm{Q}_x$, $x \in I$.*

(iii) *There exists a $\mathfrak{B}([0, \infty)) \otimes \mathcal{C}_\infty(\mathrm{Q})$-measurable functional $F : [0, \infty) \times C \to [0, \infty]$ which is right-continuous in the first variable and such that the right-inverse of T^i is just $F(t, Y^i)$:*

$$\inf\{s \geq 0 : T_s^i > t\} = F(t, Y^i), \qquad t \geq 0, \text{ a.s., } i = 1, 2.$$

Then X^1 and X^2 possess the same distribution on $\mathcal{C}_\infty^0$, i.e. $\mathbf{P}_x^1 \circ (X^1)^{-1} = \mathbf{P}_x^2 \circ (X^2)^{-1}$, $x \in I$.

Proof. Let $T_t(\omega) = \inf\{s \geq 0 : F(s, \omega) > t\}$. Combining the $\mathcal{C}_\infty(\mathrm{Q})$-measurability of $T_t(\cdot)$ and the right-continuity of $T_\cdot(\omega)$ we obtain that T is $\mathfrak{B}([0, \infty)) \otimes \mathcal{C}_\infty(\mathrm{Q})$-measurable. From (iii) we have $T_t(Y^i) = T_t^i$, $t \geq 0$, a.s., $i = 1, 2$. Now define the map $\Pi : [0, \infty) \times C \to \mathbb{R}$ by $\Pi(t, \omega) = \omega_{T_t(\omega)}$. Obviously, Π is $\mathfrak{B}([0, \infty)) \otimes \mathcal{C}_\infty(\mathrm{Q})$-measurable and it holds that

$X_t^i = \Pi(t, Y^i)$, $t \geq 0$, a.s., $i = 1, 2$. Therefore the assertion of the proposition immediately follows from (ii). □

The following proposition provides a useful result on the preservation of the strong Markov property under equality of distributions.

(1.24) Proposition (cf. [19], Th. 4 and Cor. 1) *Let $(X, \mathbb{F})$ be a strong Markov process on $(\Omega, \mathcal{F}, \mathbf{P}_x, x \in I)$. Further let $(\widetilde{X}, \widetilde{\mathbb{F}})$ denote a continuous stochastic process taking values in I defined on a family of probability spaces $(\widetilde{\Omega}, \widetilde{\mathcal{F}}, \widetilde{\mathbf{P}}_x, x \in I)$ satisfying (1.1) and (1.2) and possessing shift operators. If the distributions of X and $\widetilde{X}$ coincide, i.e. $\mathbf{P}_x \circ (X)^{-1} = \widetilde{\mathbf{P}}_x \circ (\widetilde{X})^{-1}$, $x \in I$, on C_∞^0, then $(\widetilde{X}, \mathbb{F}^{\widetilde{X}})$ is a strong Markov process as well.*

Proof. It is straightforward to show that $(\widetilde{X}, \mathbb{F}^{\widetilde{X}})$ satisfies the "simple" Markov property (1.4)(ii). It remains to prove the strong Markov property. But the equality of the distributions implies the equality of the transition probabilities and now the assertion follows by applying Corollary 1 of Theorem 4 from [19]. □

I.7 Extension of the Basic Probability Structure

Let $(\Omega, \mathcal{F}, \mathbf{P}_x, x \in I)$ be a family of probability spaces equipped with a filtration $\mathbb{F}$. Sometimes we will have to pass over to a suitable extension of this family.

(1.25) Definition A family of probability spaces $(\widetilde{\Omega}, \widetilde{\mathcal{F}}, \widetilde{\mathbf{P}}_x, x \in I)$ equipped with a filtration $\widetilde{\mathbb{F}}$ is called an *extension of* $(\Omega, \mathcal{F}, \mathbf{P}_x, x \in I)$ *and* $\mathbb{F}$ provided there exists a complete probability space $(\Omega', \mathcal{F}', \mathbf{P}')$ with a filtration $\mathbb{F}'$ such that

(i)
$$\widetilde{\Omega} = \Omega \times \Omega', \; \widetilde{\mathcal{F}} = \mathcal{F} \otimes \mathcal{F}', \; \widetilde{\mathcal{F}}_t = \mathcal{F}_t \otimes \mathcal{F}'_t, \qquad t \geq 0,$$
$$\widetilde{\mathbf{P}}_x = \mathbf{P}_x \otimes \mathbf{P}', \qquad x \in I,$$

(ii) there exists an $\mathbb{F}'$-Wiener process on $(\Omega', \mathcal{F}', \mathbf{P}')$,

(iii) $\mathcal{F}'_0$ contains all $\mathbf{P}'$-null sets of $\mathcal{F}'$.

If we pass from $(\Omega, \mathcal{F}, \mathbf{P}_x, x \in I)$, $\mathbb{F}$ to an extension $(\widetilde{\Omega}, \widetilde{\mathcal{F}}, \widetilde{\mathbf{P}}_x, x \in I)$, $\widetilde{\mathbb{F}}$ all random variables Z defined on $(\Omega, \mathcal{F})$ (resp. $(\Omega', \mathcal{F}')$) are extended to $(\widetilde{\Omega}, \widetilde{\mathcal{F}})$ in a natural way setting

$$Z(\widetilde{\omega}) = Z(\omega) \quad (\text{resp.}\, Z(\widetilde{\omega}) = Z(\omega')), \qquad \widetilde{\omega} = (\omega, \omega') \in \widetilde{\Omega},$$

without changing their notation.

If $(\Omega, \mathcal{F}, \mathbf{P}_x, x \in I)$, $\mathbb{F}$ satisfy (1.1) and (1.2), obviously the same holds true for an extension.

(1.26) Proposition *Let $(\Omega, \mathcal{F}, \mathbf{P}_x, x \in I)$, $\mathbb{F}$ satisfy (1.1) and (1.2) and consider a local martingale $Y \in \mathcal{M}(\mathbb{F}_+)$ with the quadratic variation*

$$\langle Y \rangle_t = \int_0^t f_s^2 \, ds, \qquad t \geq 0, \text{ a.s.}$$

Then on every extension $(\widetilde{\Omega}, \widetilde{\mathcal{F}}, \widetilde{\mathbf{P}}_x, x \in I)$, $\widetilde{\mathbb{F}}$ there exists a process $(B, \widetilde{\mathbb{F}}_+)$, which is a Wiener process with respect to every $\widetilde{\mathbf{P}}_x$, $x \in I$, such that

$$Y_t = \int_0^t f_s \, dB_s, \qquad t \geq 0, \text{ a.s.}$$

Proof. The assertion is essentially the well-known Proposition of Doob (see for example [29], Th. II 7.1′). Since we are working on a whole family of probability spaces we will give a construction of the desired Wiener process B.

On an extension $(\widetilde{\Omega}, \widetilde{\mathcal{F}}, \widetilde{\mathbf{P}}_x, x \in I)$, $\widetilde{\mathbb{F}}$ of $(\Omega, \mathcal{F}, \mathbf{P}_x, x \in I)$, $\mathbb{F}$, let $(B', \widetilde{\mathbb{F}}_+)$ be a Wiener process with respect to all $\widetilde{\mathbf{P}}_x$, $x \in I$. The existence of such a process is guaranteed by (1.25)(ii). Now set

$$B_t = \int_0^t f_s^{-1} \mathbf{1}_{\{f_s^2 > 0\}} \, dY_s + \int_0^t \mathbf{1}_{\{f_s^2 = 0\}} \, dB'_s, \qquad t \geq 0.$$

As mentioned above, $(\widetilde{\Omega}, \widetilde{\mathcal{F}}, \widetilde{\mathbf{P}}_x, x \in I)$, $\widetilde{\mathbb{F}}$ satisfy (1.1),(1.2) and so from (1.6)(iii) we have versions of both stochastic integrals which do not depend on the probability measure $\widetilde{\mathbf{P}}_x$ under consideration. Hence, the process B is well-defined. Finally, the well-known martingale characterization of P. Lévy yields that $(B, \widetilde{\mathbb{F}}_+)$ is a Wiener process with respect to all $\widetilde{\mathbf{P}}_x$, $x \in I$. By a straightforward computation we obtain the asserted equality for Y. □

To complete this chapter we introduce some general notation and conventions.

(1.27) For $-\infty \leq a < b \leq \infty$ let $[a, b]$ denote the real interval

$$[a,b] = \begin{cases} (-\infty, \infty) & \text{if} \quad a = -\infty, \, b = \infty, \\ [a, \infty) & \text{if} \quad a > -\infty, \, b = \infty, \\ (-\infty, b] & \text{if} \quad a = -\infty, \, b < \infty, \\ [a, b] & \text{if} \quad a, b \in \mathbb{R}. \end{cases}$$

(1.28) We use the following conventions concerning indefinite expressions:

$$0/0 = 0, \; \infty^{-1} = 0, \; 0^{-1} = \infty, \; \infty \cdot 0 = 0.$$

II. Classification of the Points of the State Space

Suppose that we are given a continuous strong Markov process $(X, \mathbb{F})$ on $(\Omega, \mathcal{F}, \mathbf{P}_x, x \in I)$ according to (1.1),(1.2) and (1.4). Recall that X takes values in the state interval I.

II.1 Classification of the Points

(2.1) For $x \in I$ we use the following abbreviated notation for certain stopping times

$$\mathrm{D}_x = \mathrm{D}_{\{x\}}(X),\ \mathrm{D}_{x+} = \mathrm{D}_{(x,\infty)\cap I}(X),\ \mathrm{D}_{x-} = \mathrm{D}_{(-\infty,x)\cap I}(X).$$

From (1.5) we have that D_x, D_{x+}, D_{x-} are $\mathbb{F}^X$-stopping times and the zero-one law of BLUMENTHAL gives $\mathbf{P}_x(\{\mathrm{D}_{x\pm} = 0\}) \in \{0,1\}$.

The following classification of the points as well as the decomposition of the state space I are well known (see [31]).

(2.2) A point $x \in I$ is called *regular* provided $\mathbf{P}_x(\{\mathrm{D}_{x+} = 0\}) = \mathbf{P}_x(\{\mathrm{D}_{x-} = 0\}) = 1$. Otherwise x is called *singular*. If all the interior points of I are regular, then the process $(X, \mathbb{F})$ itself is called *regular*, too. The singular points of I are further classified. A singular point of I is called

left singular	if	$\mathbf{P}_x(\{\mathrm{D}_{x+} = 0\}) = 0,$
right singular	if	$\mathbf{P}_x(\{\mathrm{D}_{x-} = 0\}) = 0,$
absorbing	if	$\mathbf{P}_x(\{\mathrm{D}_{x+} = 0\}) = \mathbf{P}_x(\{\mathrm{D}_{x-} = 0\}) = 0.$

Denote by R the set of regular points and by S the set of all singular points. We write K_- (resp. K_+) for the subset of left (resp. right) singular points. Finally, for the set of absorbing points $K_+ \cap K_-$ we use the symbol E.

Clearly, the set R is open (in $\mathbb{R}$) and so it is an at most countable union of disjoint intervals.

(2.3) Lemma (i) *A point $x \in I$ is left singular (resp. right singular) if and only if for all $y > x$ (resp. $y < x$) it holds that $\mathbf{P}_x(\{\mathrm{D}_y < \infty\}) = 0$.*

(ii) *In case of* $x \in (K_+ \setminus E) \cup R$ *(resp.* $x \in (K_- \setminus E) \cup R$*) for all* $y \leq x$ *(resp.* $y \geq x$*) we have that* $\mathbf{P}_y(\{\mathrm{D}_x = \mathrm{D}_{x+}\}) = 1$ *(resp.* $\mathbf{P}_y(\{\mathrm{D}_x = \mathrm{D}_{x-}\}) = 1$*).*

(iii) *It holds that* $E \in \mathfrak{B}(I)$, $X_{\mathrm{D}_E} \in E$ *on* $\{\mathrm{D}_E < \infty\}$ *a.s. and* $X = X^{\mathrm{D}_E}$.

Proof. The properties (i),(ii) are well-known and can easily be derived from the definitions above and the strong Markov property. The proof of (iii) becomes more involved because E is not closed in general. The measurability of E has already been shown in [31] (see also (2.8)). Therefore D_E is certainly an $\mathbb{F}^X$-stopping time. The last assertion in (iii) is equivalent to

$$X_{\mathrm{D}_E} = X_{\mathrm{D}_E + t} \quad \text{on} \quad \{\mathrm{D}_E < \infty\}, \qquad t \geq 0, \text{ a.s.}$$

Obviously, on the subset $\{\mathrm{D}_E < \infty, X_{\mathrm{D}_E} \in E\}$ of $\{\mathrm{D}_E < \infty\}$ the above equality follows from the strong Markov property and the definition of an absorbing point. Hence, it suffices to prove

$$\mathbf{P}_x(\{\mathrm{D}_E < \infty, X_{\mathrm{D}_E} \notin E\}) = 0, \qquad x \in I.$$

The continuity of X implies, that in case of $\mathrm{D}_E < \infty$, $X_{\mathrm{D}_E} \notin E$ the point X_{D_E} has to be a limit point of the set E. Let E^- and E^+ denote the sets $\{y \in I \setminus E : \exists (x_n), x_n \downarrow y, x \in E\}$ and $\{y \in I \setminus E : \exists (x_n), x_n \uparrow y, x \in E\}$, respectively. It is easy to verify, that every point from E^- (resp. E^+) is left (resp. right) singular. Now we have that $\mathbf{P}_x(\{\mathrm{D}_E < \infty, X_{\mathrm{D}_E} \in E^- \cap (-\infty, x)\}) = 0$ as well as $\mathbf{P}_x(\{\mathrm{D}_E < \infty, X_{\mathrm{D}_E} \in E^- \cap [x, \infty)\}) = 0$. The same is true for E^+ instead of E^- and so (iii) is proved. □

II.2 Decomposition of the State Space

In the next decomposition of I we follow [31], chapter 3.5.

(2.4) Two points $a, b \in I$ are called *connected* ("$a \sim b$") provided that there exist $x_1, ..., x_n \in I$ such that $a \leq x_1 \leq ... \leq x_n = b$ in case of $a \leq b$ and $b \leq x_1 \leq ... \leq x_n = a$ in case of $b < a$, respectively, and for every $k = 1, ..., n-1$ it holds $\mathbf{P}_{x_k}(\{\mathrm{D}_{x_{k+1}} < \infty\}) > 0$ or $\mathbf{P}_{x_{k+1}}(\{\mathrm{D}_{x_k} < \infty\}) > 0$. The relation $\sim$ is an equivalence relation which generates a decomposition of I into equivalence classes I_α, $\alpha \in J'$. The equivalence classes consisting of only one point are necessarily subsets of E. We write

$$I = E^* \cup \bigcup_{\alpha \in J} I_\alpha \quad \text{with} \quad E^* = \bigcup_{\text{card } I_\alpha = 1} I_\alpha$$

where J is chosen to be the subset of J' whose elements α satisfy $\text{card } I_\alpha > 1$. Obviously, for further investigations the set E^* does not play any role. The set J is at most countable and for each $\alpha \in J$ the class I_α is a nondegenerate interval which is called an *interval connected with respect to* X.

From the construction, we have that for every $x \in I_\alpha$ $\mathbf{P}_x(\{X_t \in I_\alpha, t \geq 0\}) = 1$ holds, that is, $(X, \mathbb{F})$ on $(\Omega, \mathcal{F}, \mathbf{P}_x, x \in I_\alpha)$ represents a strong Markov process with values in I_α[1]. Therefore, without loss of generality, we shall sometimes assume that our state interval I is connected with respect to X.

(2.5) Lemma (i) *The set K_+ (resp. K_-) is closed to the right-hand side (resp. left-hand side), that is, for $(x_n) \subseteq K_+$, $x_n \uparrow x \in I$, (resp. $(x_n) \subseteq K_-$, $x_n \downarrow x \in I$), we have that $x \in K_+$ (resp. $x \in K_-$).*

(ii) *Let I be connected with respect to X. Then the sets K_+, K_- and so $E = K_+ \cap K_-$ are closed in I. Moreover, a limit point $x \in \text{int}(I)$ of a sequence $(x_n) \subseteq K_+$, $x_1 > x_2 > \ldots > x$, (resp. $(x_n) \subseteq K_-$, $x_1 < x_2 < \ldots < x$) belongs to $K_+ \setminus E$ (resp. $K_- \setminus E$).*

Proof. Assertion (i) follows immediately from the definition of K_+ and K_-, respectively. We proceed with the proof of (ii) and start with a sequence $(x_n) \subseteq K_+$ which converges to a point x. In case of $x_n \uparrow x$ we have obviously that $x \in K_+$. Now let $x_n \downarrow x$, $x_n > x$. If $x = \inf(I) \in I$ then $x \in K_+$ is evident. So assume $x \in \text{int}(I)$[2]. Then obviously $x \in R^c = K_+ \cup K_-$. We show $x \notin K_-$ proving that $x \in K_+$. Indeed, if $x \in K_-$ were true then for all $y > x$ we would have $\mathbf{P}_x(\{\mathrm{D}_y < \infty\}) = 0$ (see (2.3)) as well as $\mathbf{P}_y(\{\mathrm{D}_x < \infty\}) = 0$. This means, that all points $y > x$ would not be connected with x, which is a contradiction.

Finally, the assertions for K_- can be proved analogously. □

As mentioned above the set R of regular points is open. Let

$$R = \bigcup_n (a_n, b_n)$$

be its representation as a countable union of disjoint open intervals. We set

$$Q = E \cup \{a_n, b_n : a_n \in K_- \setminus E,\ b_n \in K_+ \setminus E \text{ or } a_n \in K_+ \setminus E,\ b_n \in K_- \setminus E\}$$

and call the elements of Q *special singular points* (cf. [31], ch. 3.5).

(2.6) Lemma *Let I be connected with respect to X. Then the special singular points can only accumulate at the boundary points of I. Moreover, a limit point of Q is necessarily absorbing if it lies in I.*

Proof. Let be $(x_n) \subseteq Q$ with $x_n \to x \in I$. Then x is a limit point of K_+ as well as of K_- and (2.5)(ii) yields $x \in K_+ \cap K_- = E$.

Now we show that the above limit point x cannot belong to the open kernel of I. Indeed, in case of $x_n > x$ we have $\mathbf{P}_x(\{\mathrm{D}_z < \infty\}) = 0$, $\mathbf{P}_z(\{\mathrm{D}_x <$

[1] All definitions in (1.1),(1.2) then have to be done with respect to $(\mathbf{P}_x, x \in I_\alpha)$.
[2] $\text{int}(I)$ denotes the open kernel of I

$\infty\}) = 0$ for all $z > x$, while in case of $x_n < x$ the same is true for all $z < x$. But this contradicts the fact that I is connected. □

(2.7) Corollary *If I is connected with respect to X, then, in each subinterval $[a,b] \subseteq \operatorname{int}(I)$, there is at most a finite number of special singular points.*

(2.8) Lemma *The set $I \setminus E$ possesses a representation as an at most countable union of nondegenerate intervals J_n:*

$$I \setminus E = \bigcup_n J_n, \; J_n \cap J_m = \emptyset, \; n \neq m,$$

where, for all n, we have that either $J_n \setminus R \subseteq K_+ \setminus E$ or $J_n \setminus R \subseteq K_- \setminus E$ or $J_n = [a,b]$ where $(a,b) \subseteq R$, $a \in K_+ \setminus E$, $b \in K_- \setminus E$. The end-points of the intervals J_n are special singular points provided they belong to I. Moreover, all points of an interval J_n are connected.

Proof. To start with, recall that $I \setminus E \subseteq \bigcup_{\alpha \in J} I_\alpha$ where J is at most countable and every I_α is connected with respect to X (cf. (2.4)). Clearly it suffices to prove the assertion for $I = I_\alpha$.

(2.6) states that the special singular points of I can only accumulate in an absorbing boundary point of I. Therefore, the set Q of special singular points yields a decomposition of $I \setminus E$ into nondegenerate intervals I_n:

$$I \setminus E = \bigcup_n I_n, \; \operatorname{int}(I_n) \cap Q = \emptyset, \; \operatorname{int}(I_n) \cap \operatorname{int}(I_m) = \emptyset, \; n \neq m.$$

For a fixed n, let $\operatorname{int}(I_n) = (c,d)$. We set

$$J_n = (c,d) \cup \{c : c \in K_+ \setminus E\} \cup \{d : d \in K_- \setminus E\}$$

which gives

$$I \setminus E = \bigcup_n J_n, \; J_n \cap J_m = \emptyset, \; n \neq m,$$

and the asserted properties of J_n remain to be shown.

Let $x, y \in J_n$ with $x \in K_+ \setminus E$, $y \in K_- \setminus E$. First consider the case $x < y$ and define $\bar{y} = \inf\{z \in K_- : x < z \leq y\}$. Then, (2.5)(i) implies $\bar{y} \in K_- \setminus E$ and $x < \bar{y}$. Again by (2.5) for $\bar{x} = \sup\{z \in K_+ : x \leq z \leq \bar{y}\}$ we obtain $\bar{x} \in K_+ \setminus E$, $\bar{x} < \bar{y}$. By definition of $\bar{x}$ and $\bar{y}$, we have $(\bar{x}, \bar{y}) \subseteq R$ and $\bar{x}, \bar{y}$ are special singular points. This yields $J_n = [\bar{x}, \bar{y}]$.

Now we suppose $x > y$ and define $\bar{y} = \sup\{z \in K_- : x > z \geq y\}$, $\bar{x} = \inf\{z \in K_+ : \bar{y} < z \leq x\}$. As above it follows that $\bar{x}, \bar{y}$ are special singular points, $\bar{x} \in K_+ \setminus E$, $\bar{y} \in K_- \setminus E$, $x \geq \bar{x} > \bar{y} \geq y$ and $J_n = (\bar{y}, \bar{x}) \subseteq R$. But this contradicts $x, y \in J_n$.

Summarizing, we have proved the asserted alternative. Finally, the points of J_n are connected because J_n is a subinterval of I which is connected with respect to X. □

II.3 Auxiliary Results

The statements below are closely related to the classification of the points made in the last section. We will often rely on these results in the remaining part of the book.

Let $(X, \mathbb{F})$ on $(\Omega, \mathcal{F}, \mathbf{P}_x, x \in I)$ be a continuous strong Markov process according to (1.1),(1.2) and (1.4).

(2.9) Lemma *For $x \in I \setminus R$ we have that $\mathbf{P}_x(\{\mathrm{D}_z < \infty\}) \in \{0,1\}$, $\forall z \in I$. In case of $\mathbf{P}_x(\{\mathrm{D}_z < \infty\}) = 1$ it even holds that $\mathbf{E}_x \mathrm{D}_z < \infty$.*

Proof. We only consider $x \in K_+$, and the case $x \in K_-$ can be treated analogously. If $z < x$ then $\mathbf{P}_x(\{\mathrm{D}_z = \infty\}) = 1$ follows from the definition of K_+ while in the case of $x = z$ the assertion is trivial. So the case $z > x$ where $\mathbf{P}_x(\{\mathrm{D}_z < \infty\})$ is strictly positive remains to be considered. Here, there exist $t > 0$, $q > 0$ such that $\mathbf{P}_x(\{\mathrm{D}_z \leq t\}) = q$. Now, for $y \in [x, z]$ the strong Markov property gives $\mathbf{P}_y(\{\mathrm{D}_z \leq t\}) \geq \mathbf{P}_x(\{\mathrm{D}_z \leq t\})$ which implies $\mathbf{P}_y(\{\mathrm{D}_z > t\}) \leq 1 - q$ for all $y \in [x, z]$. But $x \in K_+$ yields

$$X_t \in [x, z], \qquad t < \mathrm{D}_z,\ \mathbf{P}_y\text{-a.s.},$$

for all $y \in [x, z]$. Applying the Markov property, by induction we finally obtain

$$\mathbf{P}_y(\{\mathrm{D}_z > nt\}) \leq (1 - q)^n, \qquad y \in [x, z].$$

Therefore it holds that $\mathbf{E}_x \mathrm{D}_z < \infty$ and in particular $\mathbf{P}_x(\{\mathrm{D}_z < \infty\}) = 1$. □

In the sequel, we need a finer classification of the points of K_+ and K_-. Let

$$R = \bigcup_n (a_n, b_n)$$

be the decomposition of the set of regular points into disjoint open intervals. We set

$$\begin{aligned} S_+ &= K_+ \setminus (\{a_n : a_n \in K_+\} \cup E), \\ S_- &= K_- \setminus (\{b_n : b_n \in K_-\} \cup E). \end{aligned} \tag{2.10}$$

(2.11) Lemma *Let $x \in S_+$ (resp. $x \in S_-$). Then for every $\varepsilon > 0$ we have that $(x, x + \varepsilon) \cap (K_+ \setminus E) \neq \emptyset$ (resp. $(x - \varepsilon, x) \cap (K_- \setminus E) \neq \emptyset$). Moreover, if I is connected with respect to X, then every point $x \in I$ that is a limit point of a sequence $(x_n) \subseteq S_+$, $x_n \geq x_{n+1}$, (resp. $(x_n) \subseteq S_-$, $x_n \leq x_{n+1}$) belongs to $S_+ \cup E$ (resp. $S_- \cup E$).*

Proof. Again we restrict ourselves to the proof of the assertion for S_+. Let $x \in S_+$. If, for some $\varepsilon > 0$, we have $(x, x+\varepsilon) \subseteq R$ then $x = a_n$ for some n and this implies $x \notin S_+$. Thus, for $x \in S_+$ we have that $(x, x+\varepsilon) \cap (K_+ \cup K_-) \neq \emptyset$ for all $\varepsilon > 0$. If there is a sequence $(x_n) \subseteq K_-$, $x_n \downarrow x$, then, by (2.5)(i),

$x \in K_-$, which is a contradiction. This yields $(x, x+\varepsilon) \cap K_+ \setminus E \neq \emptyset$ for all $\varepsilon > 0$.

To prove the second assertion, let $(x_n) \subseteq S_+$, $x_n \geq x_{n+1}$, tend to $x \in I$ where $x_n > x$ without loss of generality. Applying (2.5)(ii), we obtain $x \in K_+$. Finally it is impossible that $x = a_n$ for any n , so the proof is complete. □

The next lemma, which describes the first entry time into the set S_+ and S_-, respectively, will play an important role in what follows.

(2.12) Lemma *For $x \in I$ we define*

$$s_+(x) = \inf\{z \geq x : z \in S_+\},\ s_-(x) = \sup\{z \leq x : z \in S_-\}$$

where $\inf \emptyset = \infty$, $\sup \emptyset = -\infty$.

(i) *It holds that*

$$\mathrm{D}_{S_+} = \mathrm{D}_{s_+(x)} \quad on \quad \{\mathrm{D}_{S_+} < \infty\} \qquad \mathbf{P}_x - a.s.$$

and

$$\mathrm{D}_{S_-} = \mathrm{D}_{s_-(x)} \quad on \quad \{\mathrm{D}_{S_-} < \infty\} \qquad \mathbf{P}_x - a.s.$$

In case of $\mathbf{P}_x(\{\mathrm{D}_{S_+} < \infty\}) > 0$ *(resp.* $\mathbf{P}_x(\{\mathrm{D}_{S_-} < \infty\}) > 0$*) we have* $s_+(x) \in S_+$ *(resp.* $s_-(x) \in S_-$*).*

(ii) *For all* $t \geq 0$ *it holds that*

$$X_t \in I \setminus S_- \quad on \quad \{\mathrm{D}_{S_+} < \infty\} \qquad a.s.$$

and

$$X_t \in I \setminus S_+ \quad on \quad \{\mathrm{D}_{S_-} < \infty\} \qquad a.s.$$

Proof. Again, we only deal with the assertion for S_+.

(i) Let $x \in I$ be fixed. We set $\xi = \sup\{z \leq x : z \in S_+\}$. If ξ is finite we conclude from (2.5)(i) that $\xi \in K_+$. First we show

(2.13) $$\mathbf{P}_x(\{\mathrm{D}_{S_+ \cap (-\infty, x)} = \infty\}) = 1.$$

This is obvious for $\xi = x$ or $\xi = -\infty$, and, so we assume $-\infty < \xi < x$. Then $(\xi, x]$ is a subset of $I \setminus S_+$. If $(\xi, x] \cap K_+ \neq \emptyset$ assertion (2.13) follows immediately. In case of $(\xi, x] \cap K_+ = \emptyset$ we have that either $\xi = a_n \in K_+$ for some n or ξ is limit point of a sequence $(x_n) \subseteq K_-$ with $x_n \downarrow \xi$. But $\xi = a_n \in K_+$ yields $\xi \in K_+ \setminus S_+$ and (2.13) follows. In the remaining case, (2.5)(i) gives $\xi \in K_-$. Therefore, it holds that $\xi \in K_- \cap K_+ = E$ and (2.13) is also satisfied.

From (2.13) we derive $\mathrm{D}_{S_+} = \mathrm{D}_{S_+ \cap [x, \infty)}$ $\mathbf{P}_x$-a.s. In case of $s_+(x) = \infty$ there is nothing to show. Let be $s_+(x) < \infty$, then $s_+(x) \in K_+ \cup K_-$. For $s_+(x) \in K_-$ we have $\mathbf{P}_x(\{\mathrm{D}_{S_+} < \infty\}) = 0$. Therefore, if $\mathbf{P}_x(\{\mathrm{D}_{S_+} < \infty\}) > 0$ then $s_+(x) \in K_+ \setminus E$. Now $s_+(x) \notin S_+$ yields $s_+(x) = a_n$ for some n which contradicts the definition of $s_+(x)$. Consequently, $s_+(x) \in S_+$ if $\mathbf{P}_x(\{\mathrm{D}_{S_+} < \infty\}) > 0$. This proves the first assertion.

(ii) Let $x \in I$ be fixed with $\mathbf{P}_x(\{\mathrm{D}_{S_+} < \infty\}) > 0$. In view of (i) we can claim $X_t \in I \setminus S_-$, $t < \mathrm{D}_{S_+}$, $\mathbf{P}_x$-a.s., on $\{\mathrm{D}_{S_+} < \infty\}$. Hence, it remains be shown that $X_t \in I \setminus S_-$ for all $t \geq \mathrm{D}_{S_+}$ $\mathbf{P}_x$-a.s. on $\{\mathrm{D}_{S_+} < \infty\}$. In what follows we refer to $\mathbf{P}_x$-almost all $\omega \in \{\mathrm{D}_{S_+} < \infty\}$. From part (i) we have that $s_+(x) \in S_+$ which implies $X_t(\omega) \in I \cap [s_+(x), \infty)$, $t \geq \mathrm{D}_{s_+(x)}(\omega)$. If $S_- \cap [s_+(x), \infty) = \emptyset$, the proof is finished. So let $S_- \cap [s_+(x), \infty) \neq \emptyset$. Then we have $\eta := \inf\{y \geq s_+(x) : y \in S_-\} < \infty$ and from (2.5)(i) follows that $\eta \in K_-$ and $\eta > s_+(x)$. Therefore, $X_t(\omega)$ can only take values in $[s_+(x), \eta]$ for all $t \geq \mathrm{D}_{s_+(x)}(\omega)$. But the definition of η gives $[s_+(x), \eta) \cap S_- = \emptyset$ and for $\eta \notin S_-$ there is nothing to prove. Otherwise, if $\eta \in S_-$ then, by (2.11), there is a point $\tilde{\eta} \in (s_+(x), \eta) \cap (K_- \setminus E)$ leading to $X_t(\omega) \in [s_+(x), \tilde{\eta}]$ for all $t \geq \mathrm{D}_{s_+(x)}(\omega)$. This proves the assertion. □

(2.14) Lemma *The following identities hold:*

$$[X_0, X_t] \cap S_+ = [X_0, \max_{s \leq t} X_s] \cap S_+, \qquad t \geq 0, \textit{ a.s.},$$

and

$$[X_t, X_0] \cap S_- = [\min_{s \leq t} X_s, X_0] \cap S_-, \qquad t \geq 0, \textit{ a.s.}$$

Proof. Because of the symmetry it suffices to prove the first assertion which is equivalent to

$$(X_t, \max_{s \leq t} X_s] \cap S_+ = \emptyset, \qquad t \geq 0, \text{ a.s.}$$

Assume

$$\mathbf{P}_x(\{(X_t, \max_{s \leq t} X_s] \cap S_+ \neq \emptyset\}) > 0 \tag{2.15}$$

for some $t \geq 0$ and $x \in I$ from which we will generate a contradiction.

First, we observe the equality

$$\{(X_t, \max_{s \leq t} X_s] \cap S_+ \neq \emptyset\} = \bigcup_{y \in \mathbb{Q} \cap I} \{(X_t, \max_{s \leq t} X_s] \cap S_+ \cap [y, \infty) \neq \emptyset,\ y > X_t\}$$

which together with (2.15) implies that there exists a $y \in I$ such that

$$\mathbf{P}_x(\{(X_t, \max_{s \leq t} X_s] \cap S_+ \cap [y, \infty) \neq \emptyset,\ y > X_t\}) > 0.$$

But

$$\{(X_t, \max_{s \leq t} X_s] \cap S_+ \cap [y, \infty) \neq \emptyset,\ y > X_t\} \subseteq \{\mathrm{D}_{S_+ \cap [y,\infty)} < t,\ X_t < y\} \qquad \mathbf{P}_x - \text{a.s.},$$

and with $\eta = \mathrm{D}_{S_+ \cap [y,\infty)}$ we have that

$$\begin{aligned} \mathbf{P}_x(\{X_t < y,\ \eta < t\}) &\leq \mathbf{P}_x(\{\exists s : X_s < y,\ \eta < s\}) \\ &\leq \mathbf{P}_x(\{\exists s \in \mathbb{Q}_+ : X_{\eta+s} < y,\ \eta < \infty\}) \\ &\leq \sum_{s \in \mathbb{Q}} \mathbf{P}_x(\{X_{\eta+s} < y,\ \eta < \infty\}). \end{aligned}$$

Now we will see that each summand in the last sum vanishes, which is a contradiction of (2.15). Indeed, taking into account that (2.12)(i) gives

$$X_\eta \in S_+ \cap [y, \infty) \quad \text{on} \quad \{\eta < \infty\} \qquad \text{a.s.},$$

then the strong Markov property implies

$$\mathbf{P}_x(\{X_{\eta+s} < y, \, \eta < \infty\}) = \mathbf{E}_x\left(\mathbf{P}_{X_\eta}(\{X_s < y\})\mathbf{1}_{\{\eta<\infty\}}\right) = 0$$

for all $s \geq 0$. Thus we have proved

$$(X_t, \max_{s \leq t} X_s] \cap S_+ = \emptyset \qquad \text{a.s.}$$

for every $t \geq 0$, and, consequently

$$(X_t, \max_{s \leq t} X_s] \cap S_+ = \emptyset, \qquad t \in \mathbb{Q}_+, \text{ a.s.}$$

Since X is continuous this yields the assertion. □

In what follows we provide some results concerning the behaviour of X in the set R of regular points. As was mentioned already, R is open (in $\mathbb{R}$) and can be represented as an at most countable union of disjoint intervals (a_n, b_n) each of which we call a *component of R.*

(2.16) Lemma *There exists a function $p : R \to \mathbb{R}$ which is continuous and strictly increasing on each component (a, b) of R and satisfies*

$$\mathbf{P}_x(\{\mathrm{D}_y < \mathrm{D}_z\}) = \frac{p(z) - p(x)}{p(z) - p(y)} \tag{2.17}$$

for all $a < y < x < z < b$.

The reader can find a proof in [31], [28], for example.

On each component of R the function p in (2.17) above is unique up to an affine transformation.

(2.18) Definition (i) A function $p : R \to \mathbb{R}$ satisfying the assertion of Lemma (2.16) is said to be a *scale function of X.*

(ii) A boundary point c of a component (a, b) of R is called *attracting for* (a, b) provided $|\lim_{x \to c, x \in (a,b)} p(x)| < \infty$ and *nonattracting for* (a, b) otherwise. Moreover, c is called *attainable for* (a, b) if there exists an $x \in (a, b)$ such that $\mathbf{P}_x(\{\mathrm{D}_c < \infty\}) > 0$.

The assertions of the following lemma are well-known and can be easily verified.

(2.19) Lemma *Let (a,b) be a component of R.*

(i) *For all $x,y \in (a,b)$, $\mathbf{P}_x(\{\mathrm{D}_y < \infty\}) > 0$ holds.*

(ii) *For all $a < y < x < z < b$, $\mathbf{P}_x(\{\mathrm{D}_{y-} \wedge \mathrm{D}_{z+} < \infty\}) = 1$ and even $\mathbf{E}_x(\mathrm{D}_{y-} \wedge \mathrm{D}_{z+}) < \infty$ hold.*

(iii) *An attainable boundary point of (a,b) is necessarily attracting but the converse does not hold.*

(iv) *If a boundary point c of (a,b) is attainable, then it holds that $\mathbf{P}_x(\{\mathrm{D}_c < \infty\}) > 0$ for all $x \in (a,b)$.*

(2.20) Lemma (i) *Let $p : I \to \mathbb{R}$ be a function whose restriction to R is a scale function of X. Then we have $p(X) \in \mathcal{M}(\mathbb{F}, \mathrm{D}_{R^c})$.*

(ii) *Let $p : I \to \mathbb{R}$ be a function, locally bounded on R, satisfying $p(X) \in \mathcal{M}(\mathbb{F}, \mathrm{D}_{R^c})$. Then, on each component of R, p is an affine transformation of a scale function, that is, if $\widetilde{p}$ is a scale function of X, then for each component (a_n, b_n) of R there exist constants c_n, d_n such that*

$$p(x) = c_n \widetilde{p}(x) + d_n, \qquad x \in (a_n, b_n).$$

Proof. (i) Let p be a function whose restriction to R is a scale function of X. We only consider $x \in R$, because in case of $x \in R^c$ the process $(p(X), \mathbb{F})$ is trivially a continuous local $\mathbf{P}_x$-martingale up to D_{R^c}. Now, without loss of generality, choose x from a component (a,b) of R. Then $\mathbf{P}_x$-a.s. $\mathrm{D}_{R^c} = \mathrm{D}_a \wedge \mathrm{D}_b$. Now we pick $a_k \downarrow a$, $a_k > a$, $b_k \uparrow b$, $b_k < b$, and set $\tau_k = \mathrm{D}_{a_k} \wedge \mathrm{D}_{b_k}$. Then $\tau_k \uparrow \mathrm{D}_{R^c}$ $\mathbf{P}_x$-a.s. and (2.19) yields $\tau_k < \infty$ $\mathbf{P}_x$-a.s. We have to show that $(p(X^{\tau_k}), \mathbb{F})$ is a $\mathbf{P}_x$-martingale. But $p(X_{\tau_k})$ is $\mathbf{P}_x$-a.s. bounded and so the Markov property as well as (2.17) give $\mathbf{P}_x$-a.s.

$$\begin{aligned}
\mathbf{E}_x(p(X_{\tau_k})|\mathcal{F}_t) &= p(X_{\tau_k})\mathbf{1}_{\{\tau_k \le t\}} + \mathbf{E}_x(p(X_{\tau_k}) \circ \theta_t|\mathcal{F}_t)\mathbf{1}_{\{\tau_k > t\}} \\
&= p(X_{\tau_k \wedge t}),
\end{aligned}$$

which is the desired martingale property.

(ii) Assume $p(X) \in \mathcal{M}(\mathbb{F}, \mathrm{D}_{R^c})$ and let (a,b) denote an arbitrary component of R. If $p(y) = p(z)$ for two points $y, z \in (a,b)$, $y < z$, then p is constant on the whole interval (a,b) and there is nothing to prove. Indeed, for $y < x < z$ the process $(p(X^{\mathrm{D}_y \wedge \mathrm{D}_z}), \mathbb{F})$ is a bounded $\mathbf{P}_x$-martingale and due to (2.19)(ii) we have that

$$\begin{aligned}
p(x) &= \mathbf{E}_x p(X_0) \\
&= \mathbf{E}_x p(X_{\mathrm{D}_y \wedge \mathrm{D}_z}) \\
&= p(z)\mathbf{P}_x(\{\mathrm{D}_z < \mathrm{D}_y\}) + p(y)\mathbf{P}_x(\{\mathrm{D}_y < \mathrm{D}_z\}) \\
&= p(z) = p(y).
\end{aligned}$$

Analogously one shows $p(u) = p(y)$, $a < u < y$, and $p(v) = p(z)$, $z < v < b$.

So the case, where $p(y) \neq p(z)$ for all $y, z \in (a,b)$, $y \neq z$ remains to be considered. As above we derive

$$p(x) = p(z)\mathbf{P}_x(\{\mathrm{D}_z < \mathrm{D}_y\}) + p(y)\mathbf{P}_x(\{\mathrm{D}_y < \mathrm{D}_z\})$$

for all $a < y < x < z < b$ which implies (2.17) since $1 = \mathbf{P}_x(\{\mathrm{D}_z < \mathrm{D}_y\}) + \mathbf{P}_x(\{\mathrm{D}_y < \mathrm{D}_z\})$. Hence, on the interval (a,b) the function p only distinguishes from a scale function $\widetilde{p}$ of X by an affine transformation. □

(2.21) Lemma (i) *For each bounded continuous function $f : I \to \mathbb{R}$ the map $x \mapsto \mathbf{E}_x f(X_t)$ is continuous on R, $t \geq 0$.*

(ii) *Let T be a perfect terminal time (see (1.7)(ii)). If $J \subseteq R$ is an interval satisfying $\mathbf{P}_x(\{T > 0\}) = 1$ and $\mathbf{E}_x T < \infty$ for all $x \in J$, then $x \mapsto \mathbf{E}_x T$ is continuous on J.*

Proof. (i) (Compare [21], Prop. (1.9)). Fix $x \in R$ and choose a sequence $(x_n) \subseteq R$ which converges to x. Then it holds that $\mathbf{P}_x(\{\lim_{n\to\infty} \mathrm{D}_{x_n} = 0\}) = 1$. By the strong Markov property and the assumptions on f we obtain

$$\begin{aligned} \mathbf{E}_x f(X_t) &= \lim_{n\to\infty} \mathbf{E}_x \mathbf{1}_{\{\mathrm{D}_{x_n} < \infty\}} f(X_{\mathrm{D}_{x_n}+t}) \\ &= \lim_{n\to\infty} \mathbf{E}_x \mathbf{1}_{\{\mathrm{D}_{x_n} < \infty\}} \mathbf{E}_{x_n} f(X_t) \\ &= \lim_{n\to\infty} \mathbf{E}_{x_n} f(X_t). \end{aligned}$$

(ii) Let $(x_n) \subseteq J$ tend to $x \in J$. As in the proof of (i) we have $\mathbf{P}_x(\{\lim_{n\to\infty} \mathrm{D}_{x_n} = 0\}) = 1$, and, again by the strong Markov property and using the properties of T, we get

$$\begin{aligned} \mathbf{E}_x T &= \lim_{n\to\infty} \mathbf{E}_x \mathbf{1}_{\{\mathrm{D}_{x_n} < T\}} T \\ &= \lim_{n\to\infty} \mathbf{E}_x \mathbf{1}_{\{\mathrm{D}_{x_n} < T\}} (\mathrm{D}_{x_n} + \mathbf{E}_{x_n} T) \\ &= \lim_{n\to\infty} \mathbf{E}_x \mathbf{1}_{\{\mathrm{D}_{x_n} < T\}} \mathrm{D}_{x_n} + \lim_{n\to\infty} \mathbf{E}_{x_n} T\, \mathbf{P}_x(\{\mathrm{D}_{x_n} < T\}) \\ &= \lim_{n\to\infty} \mathbf{E}_{x_n} T, \end{aligned}$$

which is the desired result. □

(2.22) Lemma *We consider a component (a,b) of R satisfying $a \in K_+ \setminus E$ (resp. $b \in K_- \setminus E$). Then, for all $y \in (a,b)$ and all $x \in [a,b)$ (resp. $x \in (a,b]$), $\mathbf{E}_x \mathrm{D}_{y+} < \infty$ (resp. $\mathbf{E}_x \mathrm{D}_{y-} < \infty$) holds.*

Proof. We restrict ourselves to the case $a \in K_+ \setminus E$. It suffices to prove $\mathbf{E}_a \mathrm{D}_{y+} < \infty$. In view of $a \in K_+ \setminus E$ there has to exist a point z, $a < z < y < b$, such that $\mathbf{P}_a(\{\mathrm{D}_z < \infty\}) > 0$. According to (2.19)(i), $\mathbf{P}_z(\{\mathrm{D}_{y+} < \infty\}) > 0$ holds and, consequently, $\mathbf{P}_a(\{\mathrm{D}_{y+} < \infty\}) > 0$. Now we can proceed as in the proof of (2.9) obtaining the assertion. □

Finally we provide a lemma concerning densities of additive functionals which we shall frequently use in Chapter VII. The basic form of the lemma is due to MOTOO.

(2.23) Lemma ([11], (3.56)) *Let $(A, \mathbb{F}^X)$ be a nonnegative increasing additive process (cf. (1.7)) satisfying $A_0 = 0$ and*

$$\mathrm{d}A_t \ll \mathrm{d}t \qquad a.s.$$

Then there exists a nonnegative $\mathfrak{B}^u(I)$-measurable function $k : I \to [0, \infty)$ such that

$$A_t = \int_0^t k(X_s)\,\mathrm{d}s, \qquad t \geq 0,\ a.s.$$

III. Weakly Additive Functionals and Time Change of Strong Markov Processes

Let $(X, \mathbb{F})$ on $(\Omega, \mathcal{F}, \mathbf{P}_x, x \in I)$ be a strong Markov process according to (1.2), (1.4) and let $(T_t)_{t \geq t}$ denote an $\mathbb{F}$-time change (cf. (1.20)). We investigate the time–changed process $X \circ T$, and we are interested in conditions on the time change T ensuring that the time–changed process $X \circ T$ is again a strong Markov process. As is well-known (cf. e.g. [16]), for this to be true it suffices that the right-inverse A of T be a certain perfect additive functional (see (1.7)).

Changing time of strong Markov processes via the right-inverse of a certain perfect additive functional is a powerful tool for the construction of further strong Markov processes from a given one (see e.g. [16], [31]). In this chapter, we are going to weaken the assumption of the perfect additivity considerably. To start with, we collect some known results on the time change of strong Markov processes using perfect additive functionals.

Let $(X, \mathbb{F})$ on $(\Omega, \mathcal{F}, \mathbf{P}_x, x \in I)$ be a strong Markov process according to (1.2), (1.4) admitting shift operators $\Theta = (\theta_t)_{t \geq 0}$.

(3.1) Definition Let $(A, \mathbb{F})$ be a right–continuous increasing process taking values in $[0, \infty]$ and let ξ be an $\mathbb{F}^X$-stopping time. We call $(A, \mathbb{F})$ a *perfect additive functional of* $(X, \mathbb{F})$ *on* $[0, \xi)$ if

(i) $\mathbf{1}_{\{t<\xi\}} A_t$ is $\mathbb{F}^X_\infty$-measurable for every $t \geq 0$.

(ii) $A_s + A_t \circ \theta_s = A_{s+t}$ on $\{s + t < \xi\}$, $s, t \geq 0$, a.s.

For brevity, in the case of $\xi = \infty$ we also refer to the process $(A, \mathbb{F})$ as a *perfect additive functional of* $(X, \mathbb{F})$.

The following theorem, formulated within our special framework, is a slightly generalized version of a well-known result on the time change of a strong Markov process.

(3.2) Theorem (cf. [22], Theorem 2.26) *Let* $(X, \mathbb{F})$ *on* $(\Omega, \mathcal{F}, \mathbf{P}_x, x \in I)$ *be a continuous strong Markov process with the set* E *of absorbing points. Further, let* $(A, \mathbb{F})$ *be a perfect additive functional of* $(X, \mathbb{F})$ *on* $[0, \mathrm{D}_E)$ *and let* $T = (T_t)_{t \geq 0}$ *denote its right-inverse. We suppose that* A *satisfies the following assumptions:*

(i) *A is strictly increasing and continuous on* $[0, \mathrm{D}_E \wedge T_\infty)$,

(ii) $\{A_\infty < \infty\} \subseteq \{\lim_{t\to\infty} X_t$ *exists and belongs to* $E\}$.

We write $T_t^o = T_t \wedge \mathrm{D}_E$, $t \geq 0$. *Then, the time-changed process* $(X \circ T, \mathbb{F}^X \circ T^o)$ *on* $(\Omega, \mathcal{F}, \mathbf{P}_x, x \in I)$ *is again a continuous strong Markov process. Moreover,* $X \circ T$ *admits shift operators given by* $\tilde{\theta}_t = \theta_{T_t^o}$, $t \geq 0$. *The set*

$$M = \{x \in I : \mathbf{P}_x(\{T_\infty^o = 0\}) = 1\}$$

is the set of absorbing points for the process $X \circ T$.

As we shall see below, the condition of the perfect additivity of A and, related to this, the condition of strict increase can be weakened substantially; and, the theorem above turns out to be a special case of a more general result (see (3.6)).

Most of the results in this chapter were derived in [44] for a much wider class of Markov processes, namely the class of so-called right processes in the sense of [18] taking values in an arbitrary measurable space. However, due to the restriction to real–valued continuous strong Markov processes in our present framework we obtain some refinements of the result of [44].

Let $(A, \mathbb{F})$ be an increasing process taking values in $[0, \infty]$. We denote by

(3.3) $$H_+(A) = \{(t, \omega) : A_{t+\epsilon}(\omega) > A_t(\omega),\ \forall \epsilon > 0\}.$$

the set of *points of right-increase of* A.

(3.4) Definition Let ξ be an $\mathbb{F}^X$-stopping time. We call $(A, \mathbb{F})$ a *weakly additive functional of* $(X, \mathbb{F})$ *on* $[0, \xi)$ if

(i) $\mathbf{1}_{\{t<\xi\}} A_t$ is $\mathbb{F}_\infty^X$-measurable for every $t \geq 0$.

(ii) $A_s + A_t \circ \theta_s = A_{s+t}$ for every $(s, \omega) \in H_+(A)$, $s + t < \xi(\omega)$, a.s.

For brevity, in the case of $\xi = \infty$ the process $(A, \mathbb{F})$ is also called a *weakly additive functional of* $(X, \mathbb{F})$.

In contrast to a perfect additive functional on $[0, \xi)$, for a weakly additive functional A, the condition of additivity (ii) is only required for shifts $\theta_s(\omega)$ where (s, ω) is a point of right-increase of A. Clearly this matters only if A is not strictly increasing on $[0, \xi)$.

To see that the class of weakly additive functionals is indeed much wider than the class of (perfect) additive functionals, let us look for a first simple example of a weakly additive functional which is not (perfect) additive. In certain applications of Theorem (3.6) below, we will meet more involved examples.

(3.5) Example Let $(W, \mathbb{F})$ on $(\Omega, \mathcal{F}, \mathbf{P}_x, x \in \mathbb{R})$ be a Wiener process with shift operators $(\theta_t)_{t \geq 0}$. Let $f : \mathbb{R} \to \mathbb{R}$ denote a right–continuous increasing function. We put

$$A_t = f(\max_{u \leq t} W_u) - f(W_0), \qquad t \geq 0.$$

If f is not constant, it is easy to see that A is indeed not perfect additive. On the other hand, for $(s, \omega) \in H_+(A)$ we have the equality $f(\max_{u \leq s} W_u(\omega)) = f(W_s(\omega))$, and, consequently

$$\begin{aligned} A_s(\omega) + A_t \circ \theta_s(\omega) &= f(W_s(\omega)) - f(W_0(\omega)) \\ &\quad + f(\max_{s \leq u \leq t+s} W_u(\omega)) - f(W_s(\omega)) \\ &= f(\max_{0 \leq u \leq t+s} W_u(\omega)) - f(W_0(\omega)) \\ &= A_{t+s}(\omega). \end{aligned}$$

In other words, the process A is weakly additive.

Now we come to the mentioned earlier generalization of Theorem (3.2).

(3.6) Theorem *Let $(X, \mathbb{F})$ on $(\Omega, \mathcal{F}, \mathbf{P}_x, x \in I)$ be a continuous strong Markov process according to (1.2), (1.4) with the set E of absorbing points. Now let $(A, \mathbb{F})$ be a right–continuous process taking values in $[0, \infty]$ whose right–inverse we denote by $T = (T_t)_{t \geq 0}$. We assume that $(A, \mathbb{F})$ satisfies the following conditions:*

(i) *$(A, \mathbb{F})$ is a weakly additive functional of $(X, \mathbb{F})$ on $[0, \mathrm{D}_E)$,*

(ii) *A is continuous on $[0, \mathrm{D}_E \wedge T_\infty)$,*

(iii) *$\{A_\infty < \infty\} \subseteq \{\lim_{t \to \infty} X_t$ exists and belongs to $E\}$,*

(iv) *$A_t \circ \theta_{T_\infty} = \infty$, $t \geq 0$, a.s. on $\bigcup_n \{T_n = T_\infty < \mathrm{D}_E\}$,*

(v) *$X_{T_t -} = X_{T_t}$, $t \geq 0$, a.s.*

Put $T_t^o = T_t \wedge \mathrm{D}_E$, $t \geq 0$. Then, the time–changed process $(X \circ T, \mathbb{F}^X \circ T^o)$ on $(\Omega, \mathcal{F}, \mathbf{P}_x, x \in I)$ is a continuous strong Markov process with $\theta \circ T^o$ as shift operators.

Proof. First observe that $X \circ T$ is a.s. continuous because of (v) and (iii). Moreover, condition (v) applied for $t = 0$ implies $\mathbf{P}_x(\{(X \circ T)_0 = x\}) = 1$, $x \in I$.

In view of $X_t = X_{t \wedge \mathrm{D}_E}$ (cf. (2.3)(iii)) it suffices to consider the time change $T_t^o = T_t \wedge \mathrm{D}_E$, $t \geq 0$, which is the right inverse of

$$A_t^o = \begin{cases} A_t & , \quad t < \mathrm{D}_E \\ \infty & , \quad \text{otherwise.} \end{cases}$$

$(A^o, \mathbb{F})$ turns out to be a weakly additive functional of $(X, \mathbb{F})$ on $[0, \infty)$. Indeed, (3.4)(i) is obvious, and for $(s, \omega) \in H_+(A^o)$, we have $s < \mathrm{D}_E$, $(s, \omega) \in H_+(A)$ and we obtain

$$A_s^o(\omega) + A_t^o \circ \theta_s(\omega) = \begin{cases} A_s(\omega) + A_t \circ \theta_s(\omega) & , \quad t < \mathrm{D}_E \circ \theta_s(\omega) \\ A_s(\omega) + \infty & , \quad \text{otherwise} \end{cases}$$
$$= A_{t+s}^o(\omega),$$

where we have also used $\mathrm{D}_E \circ \theta_s = \mathrm{D}_E - s$ for $s < \mathrm{D}_E$. Because of $A_t^o = \infty$ for $t \geq \mathrm{D}_E$ we have that $T_\infty^o \leq \mathrm{D}_E \wedge T_\infty$ and in view of (ii) A^o is continuous on $[0, T_\infty^o)$.

Now let us show that $\theta \circ T^o$ are shift operators for $X \circ T = X \circ T^o$:

(3.7)
$$X_{T_t^o} \circ \theta_{T_s^o} = X_{T_s^o + T_t^o \circ \theta_{T_s^o}} = X_{T_{t+s}^o}, \; t, s \geq 0, \text{ a.s.}$$

This equality is clearly true for $T_s^o = T_\infty^o = \mathrm{D}_E$ (compare also the definition of θ_∞ in (1.3)). In the case of $T_s^o = T_\infty^o < \mathrm{D}_E$ condition (iv) implies $T_t^o \circ \theta_{T_s^o} = 0$, $t \geq 0$, a.s. and (3.7) follows. Now assume $T_s^o < T_\infty^o$. To prove (3.7) it suffices to verify

$$T_s^o + T_t^o \circ \theta_{T_s^o} = T_{t+s}^o, \; t \geq 0, \; s \geq 0, \text{ a.s.}$$

To this purpose, we use the weak additivity of A^o on $[0, \infty)$. Together with $(T_s^o(\omega), \omega) \in H_+(A)$ and the continuity of A^o on $[0, T_\infty^o)$, this yields

$$\begin{aligned} T_s^o + T_t^o \circ \theta_{T_s^o} &= T_s^o + \inf\{u \geq 0 : A_u^o \circ \theta_{T_s^o} > t\} \\ &= T_s^o + \inf\{u \geq 0 : A_{u+T_s^o}^o - A_{T_s^o}^o > t\} \\ &= T_s^o + \inf\{u \geq 0 : A_{u+T_s^o}^o > t + s\} \\ &= \inf\{u \geq T_s^o : A_u^o > t + s\} \\ &= T_{t+s}^o, \; t \geq 0, \; s \geq 0, \; T_s^o < T_\infty^o, \text{ a.s.} \end{aligned}$$

Finally, let us verify that $(X \circ T^o, \mathbb{F}^X \circ T^o)$ is indeed a strong Markov process. To this end, we first show that T^o is an $\mathbb{F}^X$-time change implying that $\mathbb{F}^X \circ T^o$ is well-defined. Since A^o is $\mathbb{F}$-adapted and A_t^o is $\mathcal{F}_\infty^X$-measurable for every t, we have $\{T_t^o \leq s\} \in \mathcal{F}_s \cap \mathcal{F}_\infty^X$, and the strong Markov property yields the equality

$$\mathbf{1}_{\{T_t^o \leq s\}} = \mathbf{P}_\mu(\{T_t^o \leq s\} | \mathcal{F}_s) = \mathbf{P}_\mu(\{T_t^o \leq s\} | \mathcal{F}_s^X) \qquad \mathbf{P}_\mu - a.s.$$

for all probabilities μ on $(I, \mathfrak{B}(I))$. This implies $\{T_t^o \leq s\} \in \mathcal{F}_s^X$, $t, s \geq 0$, and, consequently, T^o is an $\mathbb{F}^X$-time change.

Now let U be an $\mathbb{F}_+^{X \circ T^o}$-stopping time. Because of $\mathbb{F}_+^{X \circ T^o} \subseteq \mathbb{F}^X \circ T^o$ the random variable U is an $\mathbb{F}^X \circ T^o$-stopping time, too. Moreover, T_U^o is an $\mathbb{F}^X$-stopping time with $\mathcal{F}_{T_U^o}^X \supseteq \mathcal{F}_{U+}^{X \circ T^o}$. Now, by the strong Markov property of X, using condition (iii) and the definition of θ_∞ (cf. (1.3)), for every $B \in \mathfrak{B}(I)$ we obtain on $\{U < \infty\}$ $\mathbf{P}_x$-a.s.

$$\begin{aligned}
&\mathbf{P}_x\left(\{(X\circ T^o)_t\circ\theta_{T_U^o}\in B\}\mid \mathcal{F}^X_{T_U^o}\right)\\
&= \mathbf{P}_x\left(\{X_{T_t^o}\circ\theta_{T_U^o}\in B\}\mid \mathcal{F}^X_{T_U^o}\right)\mathbf{1}_{\{T_U^o<\infty\}}+\mathbf{1}_{\{X_{T_U^o}\in B,\,T_U^o=\infty\}}\\
&= \mathbf{P}_{(X\circ T^o)_U}\left(\{(X\circ T^o)_t\in B\}\right)\mathbf{1}_{\{T_U^o<\infty\}}+\mathbf{1}_{\{X_{T_U^o}\in B,\,T_U^o=\infty\}}\\
&= \mathbf{P}_{(X\circ T^o)_U}\left(\{(X\circ T^o)_t\in B\}\right).
\end{aligned}$$

For $U = t_0$ this is just the simple Markov property (1.4)(ii) for the process $(X\circ T^o, \mathbb{F}^X\circ T^o)$. Taking conditional expectations with respect to $\mathcal{F}^{X\circ T^o}_{U+}$ on both sides yields the strong Markov property (1.4)(iii) for $X\circ T$. □

Remarks. (i). Condition (v) in Theorem (3.6) was only used for the continuity of $X\circ T$. Thus, without (v) the assertion of the theorem, except for the continuity of $X\circ T$, remains valid.
(ii). In the case that A is even a perfect additive functional on $[0, \mathrm{D}_E)$ the condition (iv) can be dropped, since one can show that (iv) is a consequence of the remaining conditions (cf. [22], (2.31)).

It would be desirable to have a characterization of the set of absorbing points for the process $X\circ T$ in Theorem (3.6). In view of the corresponding assertion in Theorem (3.2) one would expect that again the set

$$M = \{x\in I : \mathbf{P}_x(\{T^o_\infty = 0\}) = 1\}$$

should be just the set of absorbing points for $X\circ T$. It is straightforward to verify that, under the conditions of Theorem (3.6), every $x\in M$ is indeed absorbing for $X\circ T$. However, as the example below shows, there can be other absorbing points for $X\circ T$. The problem of completely characterizing the set of absorbing points is still open.

(3.8) Example Let $(W, \mathbb{F})$ on $(\Omega, \mathcal{F}, \mathbf{P}_x, x\in\mathbb{R})$ be a Wiener process. Define

$$A_t = t\wedge \mathrm{D}_0(W) + L^W(t,0), \qquad t\geq 0.$$

It is easy to verify that A is a weakly additive functional of $(W,\mathbb{F})$ satisfying the assumptions of Theorem (3.6). Let T denote the right-inverse of A. For the set M introduced above, $M=\emptyset$ holds. On the other hand, the point 0 is absorbing for $W\circ T$.

To close this chapter let us illustrate Theorem (3.6) turning back to Example (3.5).

Let $(W,\mathbb{F})$ on $(\Omega,\mathcal{F},\mathbf{P}_x, x\in\mathbb{R})$ be a Wiener process and let $f:\mathbb{R}\to\mathbb{R}$ denote a continuous strictly increasing function. Again put for $t\geq 0$

$$\begin{aligned}
A_t &= f(\max_{s\leq t} W_s) - f(W_0),\\
T_t &= \inf\{s\geq 0 : A_s > t\}.
\end{aligned}$$

Assuming $\lim_{x\uparrow\infty} f(x) = \infty$, it is easy to see that A fulfills the conditions of Theorem (3.6). Thus, $(W \circ T, \mathbb{F}^W \circ T)$ on $(\Omega, \mathcal{F}, \mathbf{P}_x, x \in \mathbb{R})$ is a continuous strong Markov process whose structure can be discovered explicitly. Indeed, set

$$g(x) = \inf\{y \in \mathbb{R} : f(y) > x\}, \qquad x \in (f(-\infty), \infty).$$

Then, using the continuity of f we obtain

$$\begin{aligned} T_t &= \inf\{s \geq 0 : A_s > t\} \\ &= \inf\{s \geq 0 : f(\max_{u \leq s} W_u) > t + f(W_0)\} \\ &= \inf\{s \geq 0 : W_s > g(t + f(W_0))\}, \end{aligned}$$

and, consequently,

$$(W \circ T)_t = W_{T_t} = g(t + f(W_0)), \qquad t \geq 0, \ a.s.$$

In other words, the continuous strong Markov process $W \circ T$ admits increasing paths and is deterministic up to its initial value. We will return to this topic in Section IV.1.

IV. Semimartingale Decomposition of Continuous Strong Markov Semimartingales

IV.1 Continuous Strong Markov Processes with Increasing Paths

In this section we provide a remarkable result, due to CINLAR [9], on the structure of continuous strong Markov processes admitting increasing paths. Roughly speaking, the result states that every real-valued continuous strong Markov process with increasing paths is deterministic up to its initial value. This result is of fundamental importance in what follows.

(4.1) Proposition *(cf. [9]) Let $(X, \mathbb{F})$ on $(\Omega, \mathcal{F}, \mathbf{P}_x, x \in I)$ be a continuous strong Markov process according to (1.2), (1.4) and suppose that the paths $X_\cdot(\omega)$, $\omega \in \Omega$, are all increasing functions from $\mathbb{R}_+$ into the state interval I. Then there exits a $\mathfrak{B}^u(I) \otimes \mathfrak{B}(\mathbb{R}_+)$-measurable function $h : I \times \mathbb{R}_+ \to I$ such that*

$$X_t = h(X_0, t), \qquad t \geq 0, \text{ a.s.}$$

Moreover, the function h satisfies $h(h(x,t),s) = h(x, t+s)$, $x \in I$, $t, s \geq 0$, and $h(x, \cdot)$ is continuous and strictly increasing on $[0, \mathrm{D}_E(h(x,\cdot)))$.

Proof. Without restricting the generality, we assume that I is connected with respect to X (cf. (2.4)). Since the paths of X are increasing, this means

$$\mathbf{P}_x(\{\mathrm{D}_y < \infty\}) > 0, \text{ for every } x < y, \; x, y \in I.$$

Further, observe that then at most $\sup I$ can be an absorbing point for X. Let

$$A_t = X_t - X_0, \qquad t \geq 0, \text{ a.s.}$$

Then A is perfect additive, continuous and strictly increasing on $[0, \mathrm{D}_E)$ with $A = A^{\mathrm{D}_E}$. Now, let T denote the right-inverse of A and put $\hat{\theta}_t = \theta_{T_t}$, $t \geq 0$. We have $T_t = \infty$ for $t \geq A_\infty = \sup I - X_0$ and $T_t < \infty$ for $t < A_\infty$. From the perfect additivity of A it follows that

$$T_s + T_t \circ \hat{\theta}_s = T_{t+s}.$$

The strong Markov property now yields $\mathbf{P}_x$-a.s. on $\{T_s < \infty\}$

$$\mathbf{P}_x\left(\{T_{t+s} - T_s \in B\} | \mathcal{F}^X_{T_s}\right) = \mathbf{P}_{X_{T_s}}(\{T_t \in B\}) = \mathbf{P}_{s+x}(\{T_t \in B\})$$

for every $B \in \mathfrak{B}(\mathbb{R}_+)$. In other words, with respect to $\mathbf{P}_x$ the process $(T_t)_{0\le t<\sup I - x}$ is a process with independent increments. Moreover, T is continuous and strictly increasing on $[0, \sup I - x)$ $\mathbf{P}_x$-a.s. Consequently, the process T is deterministic $\mathbf{P}_x$-a.s., i.e. for every $x \in I$ there exists a function $g(x,\cdot)|\mathbb{R}_+ \to [0,\infty]$ such that

$$T_t = g(x,t), \qquad t \ge 0,\ \mathbf{P}_x - a.s.$$

From this we obtain w.r.t. $\mathbf{P}_x$

$$\begin{aligned} X_t &= A_t + X_0 \\ &= \inf\{s \ge 0 : g(x,s) > t\} + x \\ &=: h(x,t). \end{aligned}$$

Obviously, $h(x,\cdot)$ is continuous and strictly increasing on $[0, \tau(x))$ where $\tau(x) = \lim_{t\uparrow \sup I - x} g(x,t) + x = \mathrm{D}_E(h(x,\cdot))$ because $\sup I$ is the only possible absorbing point. In view of $\mathbf{E}_x X_t = h(x,t)$ the function h is a $\mathfrak{B}^u(I)\otimes\mathfrak{B}(\mathbb{R}_+)$-measurable mapping from $I \times \mathbb{R}_+$ into I. Finally, the Markov property of X yields the relation $h(h(x,t),s) = h(x,t+s)$. □

IV.2 Intrinsic Structure of the Finite Variation Part in the Semimartingale Decomposition

Let $(X,\mathbb{F})$ on $(\Omega,\mathcal{F},\mathbf{P}_x, x\in I)$ be a continuous strong Markov semimartingale with the decomposition

$$X = X_0 + M(X) + V(X),\ \ M(X) \in \mathcal{M}(\mathbb{F}^X),\ V(X) \in \mathcal{V}(\mathbb{F}^X),$$

according to (1.10). In this section, we investigate the intrinsic structure of the finite variation part $V(X)$. In Theorem (4.3) we express the process $V(X)$ explicitly using the local time of X and the process X itself. To begin, we first need the following lemma.

(4.2) Lemma *Let $(X,\mathbb{F})$ on $(\Omega,\mathcal{F},\mathbf{P}_x, x \in I)$ be a continuous strong Markov process and let p denote a scale function for X. Let (a,b) be a component of R and consider an interval $(c,d) \subseteq (a,b)$ with $\lim_{x\downarrow c} p(x) > -\infty$, $\lim_{x\uparrow d} p(x) < \infty$. In the case of $c = a$ (resp. $d = b$), we assume that p is extended in its domain to the point c (resp. d) by $p(c) = \lim_{x\downarrow c} p(x)$ (resp. $p(d) = \lim_{x\uparrow d} p(x)$).*

(i) *Denote $Y = p((X\vee c)\wedge d)$. Then Y is a continuous semimartingale, i.e. $Y \in \mathcal{S}(\mathbb{F}^X)$, and Y admits the decomposition*

$$Y_t = Y_0 + M_t + L^Y(t,p(c)) - L^Y(t,p(d)),\ \ M \in \mathcal{M}(\mathbb{F}^X).$$

(ii) *If moreover, $X \in \mathcal{S}(\mathbb{F}^X)$, i.e. X is a continuous strong Markov semimartingale, then the inverse $q : (p(a+), p(b-)) \to \mathbb{R}$ of the scale p (restricted to (a,b)) is a difference of convex functions on $(p(a+), p(b-))$. For $a \in K_+ \setminus E$ and $p(a+) > -\infty$ (resp. $b \in K_- \setminus E$ and $p(b-) < \infty$) the inverse q of p restricted to $[a,b)$ (resp. $(a,b]$) is even a difference of convex functions on $[p(a+), p(b-))$ (resp. $(p(a+), p(b-)]$).*

Proof. (i). We utilize an idea from the proof of Thèoréme 3 in [38]. Put

$$\begin{aligned} \xi_t &= \inf\{s \geq t : X_s \notin (c,d)\} \\ &= \mathrm{D}_{(c,d)^c} \circ \theta_t + t, \ t \geq 0. \end{aligned}$$

Then, applying the strong Markov property of X, for every $\mathbb{F}^X$-stopping time T we obtain $\mathbf{P}_x$-a.s.

$$\begin{aligned} \mathbf{E}_x\left(Y_{\xi_T} | \mathcal{F}_T^X\right) \mathbf{1}_{\{T<\infty\}} &= \mathbf{E}_{X_T}\Big(p\left((X_{\mathrm{D}_{(c,d)^c}} \vee c) \wedge d\right)\Big) \mathbf{1}_{\{T<\infty\}} \\ &= \Big\{ Y_T \mathbf{1}_{\{X_T \notin (c,d)\}} + \Big[p(c) \frac{p(d) - p(X_T)}{p(d) - p(c)} + \\ &\qquad p(d) \frac{p(X_T) - p(c)}{p(d) - p(c)} \Big] \mathbf{1}_{\{X_T \in (c,d)\}} \Big\} \mathbf{1}_{\{T<\infty\}} \\ &= Y_T \mathbf{1}_{\{T<\infty\}}, \end{aligned}$$

where we have also used (2.17), (2.19)(ii). The identity just derived says that Y is nothing else but the optional projection of the process $(Y_{\xi_t})_{t\geq 0}$ (cf. [12]). Now let

$$\begin{aligned} T_0 &= \mathrm{D}_{p(c)}(Y) \wedge \mathrm{D}_{p(d)}(Y), \\ T_{n+1} &= \inf\left\{ s \geq T_n : |Y_s - Y_{T_n}| \mathbf{1}_{\{T_n<\infty\}} = p(d) - p(c) \right\}, \\ & \quad n = 0, 1, \ldots \end{aligned}$$

Every T_n is an $\mathbb{F}^X$-stopping time; and $T_n \uparrow \infty$ holds. Now we have $\mathbf{P}_x$-a.s.

$$\begin{aligned} Y_{t\wedge T_n} &= Y_t \mathbf{1}_{\{t \leq T_n\}} + Y_{T_n} \mathbf{1}_{\{t > T_n\}} \\ &= \mathbf{E}_x\left(Y_{\xi_t} | \mathcal{F}_t^X\right) \mathbf{1}_{\{t \leq T_n\}} + Y_{T_n} \mathbf{1}_{\{t > T_n\}} \\ &= \mathbf{E}_x\left(\sum_{k=0}^{\infty} Y_{T_k} \mathbf{1}_{\{T_k \leq \xi_t < T_{k+1}\}} \Big| \mathcal{F}_t^X \right) \mathbf{1}_{\{t \leq T_n\}} + Y_{T_n} \mathbf{1}_{\{t > T_n\}} \\ &= \mathbf{E}_x\left(\sum_{k=0}^{\infty} Y_{T_k} \mathbf{1}_{\{T_k \leq \xi_t\}} \Big| \mathcal{F}_t^X \right) \mathbf{1}_{\{t \leq T_n\}} \\ &\quad - \mathbf{E}_x\left(\sum_{k=0}^{\infty} Y_{T_k} \mathbf{1}_{\{T_{k+1} \leq \xi_t\}} \Big| \mathcal{F}_t^X \right) \mathbf{1}_{\{t \leq T_n\}} + Y_{T_n} \mathbf{1}_{\{t > T_n\}}. \end{aligned}$$

Assuming $p(c) > 0$ without restricting the generality, both conditional expectations are integrable submartingales, and, consequently, $Y_{\cdot\wedge T_n}$ turns out to be a $\mathbf{P}_x$-semimartingale. In view of $T_n \uparrow \infty$ this yields $Y \in \mathcal{S}(\mathbb{F}^X)$.

Now let

$$Y = Y_0 + M + V$$

denote the semimartingale decomposition of Y, $M \in \mathcal{M}(\mathbb{F}^X)$, $V \in \mathcal{V}(\mathbb{F}^X)$ (cf. (1.6)(i)). To prove the second assertion in (4.2)(i), we first show that $\mathbf{1}_{(p(c),p(d))}(Y) * Y \in \mathcal{M}(\mathbb{F}^X)$. The equality $\mathbf{E}_x(Y_{\xi_t}|\mathcal{F}_t^X) = Y_t$ implies

$$\int_t^{\xi_t} |\mathrm{d}V_s| = 0 \qquad \mathbf{P}_x\text{-a.s.}$$

Because of [1]

$$\bigcup_{r\in\mathbb{Q}_+}]\!]r,\xi_r[\![= \{(t,\omega) : t > 0,\, Y_t(\omega) \in (p(c),p(d))\}$$

from this we obtain $\int_0^\cdot \mathbf{1}_{(p(c),p(d))}(Y_s)\,|\mathrm{d}V_s| = 0$, and, thus $\mathbf{1}_{(p(c),p(d))}(Y) * Y \in \mathcal{M}(\mathbb{F}^X)$. Now, in view of $p(c) \le Y \le p(d)$, and applying the generalized Itô-formula to $(Y-p(c))^+$ and $(p(d)-Y)^+$, we get $\int_0^t \mathbf{1}_{\{p(c)\}}(Y_s)\,\mathrm{d}V_s = L^Y(t,p(c))$ and $\int_0^t \mathbf{1}_{\{p(d)\}}(Y_s)\,\mathrm{d}V_s = -L^Y(t,p(d))$. This ends the proof of assertion (i).

(ii). In the case of $a \in K_+ \setminus E$ and $p(a+) > -\infty$, put $c = a$, otherwise let c be arbitrary with $a < c < b$. Analogously, set $b = d$ for $b \in K_- \setminus E$ and $p(b-) < \infty$, and, otherwise $c < b < d$ arbitrarily. Changing time in the process Y from part (i) using the time change $T_t = \inf\{s \ge 0 : \langle Y\rangle_s > t\}$ gives us a process $W = Y\circ T$, $\mathbb{G} = \mathbb{F}\circ T$ that is, according to (A1.5), a Wiener process reflected in $[c,d]$ and stopped at $\langle Y\rangle_\infty$. Since X is a continuous strong Markov semimartingale, we have $q(Y) = (X\vee c)\wedge d \in \mathcal{S}(\mathbb{F}^X)$. Changing time by $T^{\langle Y\rangle_n} = (T_t^{\langle Y\rangle_n})_{t\ge0}$, we obtain $q(W^{\langle Y\rangle_n}) \in \mathcal{S}(\mathbb{F}^X \circ T)$ for every n (cf. (1.21)); in other words, q is a semimartingale function for $W^{\langle Y\rangle_n}$, $n = 1,2,\ldots$

Now let $x \in (\tilde{c},\tilde{d}) \subset (c,d)$. In view of (2.19), $\mathbf{P}_x(\{\mathrm{D}_{p(\tilde{c})}(Y) \wedge \mathrm{D}_{p(\tilde{d})}(Y) < \infty\}) = 1$, and thus

$$\mathbf{P}_x\Big(\{\mathrm{D}_{p(\tilde{c})}(W) \wedge \mathrm{D}_{p(\tilde{d})}(W) < \langle Y\rangle_\infty\}\Big) = 1.$$

Consequently, defining $\tau = \mathrm{D}_{p(\tilde{c})}(W) \wedge \mathrm{D}_{p(\tilde{d})}(W)$, the process $q(W^\tau)$ is a $\mathbf{P}_x$-semimartingale. According to (A1.6)(iii), this yields that the function q is a difference of convex functions on $(p(\tilde{c}),p(\tilde{d}))$, and, consequently, on $(p(a),p(b))$ as well.

Finally, let us show that the assertion concerning the function q can be strengthened for $c = a \in K_+ \setminus E$, $p(a+) > -\infty$. For $b \in K_- \setminus E$, $p(b-) < \infty$ the corresponding assertion is proved analogously. By (2.22), with respect to

[1] $]\!]S,T[\![= \{(t,\omega) : S(\omega) < t < T(\omega)\}$ (cf. [12]).

$\mathbf{P}_a$ we have $\mathbf{P}_a(\{\mathrm{D}_y(X) < \infty\}) = 1$ for every $a \le y < d$. This in turn implies $\mathbf{P}_a(\{\mathrm{D}_{p(y)}(Y) < \infty\}) = 1$ and thus

$$\mathbf{P}_a\left(\{\langle Y\rangle_\infty \ge \mathrm{D}_{p(y)}(W)\}\right) = \mathbf{P}_a\left(\{\langle Y\rangle_\infty \ge \mathrm{D}_{p(d)}(W)\}\right) = 1.$$

Denoting $\tau = \mathrm{D}_{p(d)}(W)$, this yields that $q(W^\tau)$ is a semimartingale under $\mathbf{P}_a$. Then (A1.6)(iii) implies that q is a difference of convex functions on the interval $[p(a), p(d))$. □

Now we come to the main result of this chapter.

(4.3) Theorem *Let $(X, \mathbb{F})$ on $(\Omega, \mathcal{F}, \mathbf{P}_x, x \in I)$ be a continuous strong Markov semimartingale with the decomposition*

$$X = X_0 + M(X) + V(X),\ M(X) \in \mathcal{M}(\mathbb{F}^X),\ V(X) \in \mathcal{V}(\mathbb{F}^X).$$

Denote by

$$R = \bigcup_n (a_n, b_n)$$

the representation of the set R of all regular points as an at most countable union of its components. Moreover, let (cf. (2.10))

$$S_+ = K_+ \setminus (\{a_n : a_n \in K_+\} \cup E),\ S_- = K_- \setminus (\{b_n : b_n \in K_-\} \cup E).$$

For a given scale function $p : R \to \mathbb{R}$ we denote by $q_n : (p(a_n+), p(b_n-)) \to (a_n, b_n)$ the inverse of the function p restricted to (a_n, b_n), $n = 1, 2, \ldots$ We define the function $g : R \to [0, \infty)$ by

$$g(x) = (q_n)'_+(p(x)),\ x \in (a_n, b_n),\ n = 1, 2, \ldots \tag{4.4}$$

Then there exist $\mathfrak{B}^u(I) \otimes \mathfrak{B}(\mathbb{R}_+)$-measurable functions

$$d_+ : I \times \mathbb{R}_+ \to [0, \infty],\ d_- : I \times \mathbb{R}_+ \to [0, \infty],$$

continuously increasing in the second variable, such that

$$\begin{aligned} V_t(X) &= \frac{1}{2}\int_R L^X(t, y)\mu_g(\mathrm{d}y) + \sum_{n:\, a_n \in K_+} L^X(t, a_n) - \sum_{n:\, b_n \in K_-} L^X(t, b_n) \\ &\quad + d_+\Big(X_0, \int_0^t \mathbf{1}_{S_+}(X_s)\mathrm{d}s\Big) - d_-\Big(X_0, \int_0^t \mathbf{1}_{S_-}(X_s)\mathrm{d}s\Big), \end{aligned}$$

$t \ge 0$, a.s., where the integral w.r.t. μ_g is defined in the Appendix 2 (see (A2.7)).

We divide the proof of the theorem into three lemmas, each describing the behaviour of the process $V(X)$ on the set R, $\{a_n\}$ with $a_n \in K_+$ (resp. $\{b_n\}$ with $b_n \in K_-$) and on S_+ (resp. S_-).

We always suppose that the assumptions of Theorem (4.3) are fulfilled.

(4.5) Lemma *Let the function g be defined by (4.4), then*

$$\int_0^t \mathbf{1}_R(X_s)\mathrm{d}V_s(X) = \frac{1}{2}\int_R L^X(t,y)\mu_g(\mathrm{d}y), \qquad t \geq 0, \ a.s.$$

Proof. It suffices to show that, for a component (a,b) of R and for $[c,d] \subset (a,b)$, the equality

$$\int_0^t \mathbf{1}_{(c,d)}(X_s)\mathrm{d}V_s(X) = \frac{1}{2}\int_{(c,d)} L^X(t,y)(\bar{g}(y))^{-1}\mathrm{d}g(y), \qquad t \geq 0, \ a.s.$$

holds true (cf. (A2.8)), were $\bar{g}(y) = (g(y)+g(y-))/2$, $y \in (a,b)$. As in Lemma (4.2)(i) let $Y = p((X \vee c) \wedge d)$ and let q denote the inverse of p restricted to (a,b). In view of (4.2)(ii) the derivative q'_+ is well-defined and the function $g = q'_+ \circ p$ is locally of bounded variation on (a,b). Thus, applying the generalized Itô-formula (1.14) for $q(Y) = (X \vee c) \wedge d$ we get

$$(X_t \vee c) \wedge d = q(Y_t) = X_0 + \int_0^t q'(Y_s)\mathrm{d}Y_s + \frac{1}{2}\int L^Y(t,y)\mathrm{d}q'_+(y), \ \ t \geq 0, \ a.s.$$

From this we obtain

$$\begin{aligned}
&\int_0^t \mathbf{1}_{(c,d)}(X_s)\mathrm{d}X_s \\
&= \int_0^t \mathbf{1}_{(c,d)}(X_s)\mathrm{d}M_s(X) + \int_0^t \mathbf{1}_{(c,d)}(X_s)\mathrm{d}V_s(X) \\
&= \int_0^t \mathbf{1}_{(p(c),p(d))}(Y_s)q'(Y_s)\mathrm{d}Y_s + \frac{1}{2}\int_{(p(c),p(d))} L^Y(t,y)\mathrm{d}q'_+(y).
\end{aligned}$$

In view of the particular form (4.2)(i) of the semimartingale decomposition of Y this implies

$$\int_0^t \mathbf{1}_{(c,d)}(X_s)\mathrm{d}V_s(X) = \frac{1}{2}\int_{(p(c),p(d))} L^Y(t,y)\mathrm{d}q'_+(y), \qquad t \geq 0, \ a.s.$$

Our aim is now to express the local time of Y by the local time of X. To this end we use (1.18), (1.19). Applying Theorem 2 of [51] and (4.2)(i) we have that the local time $L^Y(t,\cdot)$ is continuous on $(p(c),p(d))$. Consequently, (1.18) yields

$$L^X(\cdot,q(y)) = L^Y(\cdot,y)q'(y), \qquad y \in (p(c),p(d)), \ a.s.$$

Thus, for every $y \in (p(c),p(d))$ with $q'(y) \neq 0$ we can replace $L^Y(t,y)$ by $L^X(t,q(y))(q'(y))^{-1}$ to get

$$\begin{aligned}
\int_0^t \mathbf{1}_{(c,d)}(X_s)\mathrm{d}V_s(X) &= \frac{1}{2}\int_{(p(c),p(d))\cap\{q'\neq 0\}} L^X(t,q(y))(q'(y))^{-1}\mathrm{d}q'_+(y) \\
&\quad + \frac{1}{2}\int_{(p(c),p(d))\cap\{q'=0\}} L^Y(t,y)\mathrm{d}q'_+(y).
\end{aligned}$$

Finally, by (A2.3)(ii) the integral $\int_{\{q'=0\}} |dq'_+|$ vanishes proving the assertion. □

Based on the proof of the preceding lemma we derive a useful characterization of those points where the local time of X vanishes.

(4.6) Let the function g on R be defined by (4.4). For every $x \in R$ we set $\bar{g}(x) = 1/2\,(g(x) + g(x-))$. We extend g in its domain of definition to the set

$$\bar{R} = R \cup \{a_n : a_n \in K_+\} \cup \{b_n : b_n \in K_-\}$$

as follows. For every component (a,b) of R with $a \in K_+$ (resp. $b \in K_-$) we put

$$\bar{g}(a) = \begin{cases} \lim_{x\downarrow a} g(x), & \text{if } a \in K_+ \setminus E,\ p(a+) > -\infty \\ 0 \quad , & \text{otherwise} \end{cases}$$

$$\left(\text{resp.}\quad \bar{g}(b) = \begin{cases} \lim_{x\uparrow b} g(x), & \text{if } b \in K_- \setminus E,\ p(b-) < \infty \\ 0 \quad , & \text{otherwise} \end{cases}\right)$$

where the existence of the respective limits follows from (4.2)(ii).

(4.7) Lemma *For every $y \in \bar{R}$, $L^X(t,y) = 0$, $t \geq 0$, holds a.s. if and only if $\bar{g}(y) = 0$. For $\bar{g}(y) \neq 0$ we even have the stronger relation $L^X(t,y) > 0$, $t > 0$, $\mathbf{P}_y$-a.s.*

Proof. First, we show that the assertion holds for $y \in R$. So let $y \in (c,d)$ where $[c,d] \subseteq (a,b)$ for some component (a,b) of R. We use the notation from the proof of (4.5). There we have derived that

$$L^X(\cdot,y) = L^Y(\cdot,p(y))\bar{g}(y), \qquad y \in (c,d),\ a.s.$$

For $\bar{g}(y) = 0$ it follows that $L^X(\cdot,y) = 0$, a.s. Now let us show that for $y \in (c,d)$ we have $\mathbf{P}_y(\{L^Y(t,p(y)) > 0\}) = 1$, for every $t > 0$. Then, by the identity above, this yields the assertion.

Since y is a regular point for X, using the definition of Y and its semimartingale decomposition (cf. (4.2)(i)), we have $\mathbf{P}_y(\{\langle M\rangle_t > 0\}) = 1$, where M is the local martingale part of Y. Now we can apply (1.17)(vi) to get $\mathbf{P}_y(\{L^Y(t,p(y)) > 0\}) = 1,\ t > 0$.

Finally, let us show the assertions for $y = a \in K_+$, where (a,b) is some component of R.

For $a \in E$ we have $\bar{g}(a) = 0$; on the other hand, by (1.17) $L^X(\cdot,a) = 0$, *a.s.*

In the case of $a \in K_+ \setminus E$, $p(a+) = -\infty$, (2.19)(iii) implies $\mathbf{P}_x(\{D_a < \infty\}) = 0$, $x > a$. Using (1.17)(v), from this we obtain $L^X_+(\cdot,a) = 0$. On the other hand, (1.17)(iv) and $a \in K_+$ give us $L^X_-(\cdot,a) = 0$, and thus $L^X(\cdot,a) = 0$.

Now let $a \in K_+ \setminus E$ and $p(a+) > -\infty$. As in (4.2) we set $Y = p((X \vee a) \wedge d)$ with some fixed d and $a < d < b$. Taking into account (4.2)(ii) and applying (1.18) with q as the inverse of p restricted to $[a,d)$, we obtain

$$L_+^X(\cdot,y) = L_+^Y(\cdot,p(y))q'_+(p(y)), \qquad a.s.,\ a < y < d.$$

Thus, using (1.12) for $y \downarrow a$, we have a.s.

$$\begin{aligned} L_+^X(\cdot,a) &= L_+^Y(\cdot,p(a+))q'_+(p(a+)) \\ &= L_+^Y(\cdot,p(a+))\bar{g}(a) \end{aligned}$$

In view of $L_-^Y(\cdot,p(a+)) = L_-^X(\cdot,a) = 0$ (cf. (1.17)(iv)) it follows that

$$L^X(\cdot,a) = \frac{1}{2}L_+^Y(\cdot,p(a+))\bar{g}(a).$$

As above in the case of $y \in R$, applying (1.17)(vi) we can show that $L_+^Y(t,p(a+)) > 0$, $t > 0$, $\mathbf{P}_a$-a.s. and from this the assertion follows. □

(4.8) Lemma *For $a_n \in K_+$ it holds that*

$$\int_0^t \mathbf{1}_{\{a_n\}}(X_s)\mathrm{d}V_s(X) = L^X(t,a_n), \qquad t \geq 0,\ a.s.$$

For $b_n \in K_-$ it holds that

$$\int_0^t \mathbf{1}_{\{b_n\}}(X_s)\mathrm{d}V_s(X) = -L^X(t,b_n), \qquad t \geq 0,\ a.s.$$

Proof. Since the proof for $b_n \in K_-$ is analogous, we only show the assertion for $a_n \in K_+$.

Abbreviate

$$k_t = \int_0^t \mathbf{1}_{\{a_n\}}(X_s)\mathrm{d}V_s(X), \qquad t \geq 0.$$

For $t \leq \mathrm{D}_{a_n}$ we have the obvious equality $k_t = 0 = L^X(t,a_n)$. It remains to show that

$$k_{\mathrm{D}_{a_n}+t} = L^X(\mathrm{D}_{a_n}+t,a_n), \qquad t \geq 0,$$

a.s. on $\{\mathrm{D}_{a_n} < \infty\}$. Using the strong Markov property and the perfect additivity of $L^X(\cdot,a_n)$ and k (see (1.16), (1.8)), we get

$$\mathbf{P}_x\Big(\{k_{\mathrm{D}_{a_n}+t} = L^X(\mathrm{D}_{a_n}+t,a_n)\} \cap \{\mathrm{D}_{a_n} < \infty\}\Big) = \mathbf{P}_{a_n}\Big(\{k_t = L^X(t,a_n)\}\Big)\mathbf{P}_x(\{\mathrm{D}_{a_n} < \infty\}).$$

Therefore, it remains to be verified that $\mathbf{P}_{a_n}\big(k_t = L^X(t,a_n)\big) = 1$, $t \geq 0$. Because of $a_n \in K_+$ we have $X \geq a_n$ w.r.t. $\mathbf{P}_{a_n}$ and thus $X = (X-a_n)^+ + a_n$. Applying the generalized Itô-formula (1.14) for $(X-a)^+$ from this we obtain

$$\int_0^t \mathbf{1}_{\{a_n\}}(X_s)\mathrm{d}V_s(X) = \frac{1}{2}L_+^X(t,a_n), \qquad t \geq 0,\ \mathbf{P}_{a_n}-a.s.$$

Because of $a_n \in K_+$, we further have $L_-^X(\cdot,a_n) = 0$ (cf. (1.17)(iv)) which implies the assertion. □

(4.9) Lemma *For every $x \in I$ there exist continuous increasing functions $d_+(x,\cdot) : \mathbb{R}_+ \longrightarrow [0,\infty]$, $d_-(x,\cdot) : \mathbb{R}_+ \longrightarrow [0,\infty]$ such that*

$$(4.10)\qquad \begin{aligned} \int_0^t \mathbf{1}_{S_+}(X_s)\mathrm{d}V_s(X) &= d_+\left(x, \int_0^t \mathbf{1}_{S_+}(X_s)\mathrm{d}s\right), \ t \geq 0,\ \mathbf{P}_x\text{-}a.s. \\ \int_0^t \mathbf{1}_{S_-}(X_s)\mathrm{d}V_s(X) &= d_-\left(x, \int_0^t \mathbf{1}_{S_-}(X_s)\mathrm{d}s\right), \ t \geq 0,\ \mathbf{P}_x\text{-}a.s. \end{aligned}$$

Proof. We restrict ourselves to the proof of the assertion for S_+.

First, it suffices to show that for every $x \in S_+$ there is a continuous increasing function $d_+(x,\cdot) : \mathbb{R}_+ \longrightarrow [0,\infty]$ with the desired properties. Then, if we define $d_+(x,\cdot)$ for $x \in I$ by

$$d_+(x,\cdot) = \begin{cases} d_+(s_+(x),\cdot) & : \ s_+(x) \in S_+ \\ 0 & : \ \text{otherwise,} \end{cases}$$

where $s_+(x) = \inf\{y \geq x : \ y \in S_+\}$ (cf. (2.12)), the assertion of the Lemma is obtained as follows. On $\{t < \mathrm{D}_{S_+}\}$ the assertion follows from

$$d_+(x,0) = 0 = \int_0^t \mathbf{1}_{S_+}(X_s)\mathrm{d}V_s(X) \qquad \mathbf{P}_x - a.s.$$

Therefore, on $\{\mathrm{D}_{S_+} < \infty\}$ it remains to show that for every $t \geq 0$ $\mathbf{P}_x$-a.s.

$$(4.11)\qquad \int_0^{\mathrm{D}_{S_+}+t} \mathbf{1}_{S_+}(X_s)\mathrm{d}V_s(X) = d_+\left(x, \int_0^{\mathrm{D}_{S_+}+t} \mathbf{1}_{S_+}(X_s)\mathrm{d}s\right).$$

In the case of $\mathbf{P}_x(\{\mathrm{D}_{S_+} < \infty\}) > 0$, by (2.12)(i) we have on $\{\mathrm{D}_{S_+} < \infty\}$ $\mathbf{P}_x$-a.s. the relation $\mathrm{D}_{S_+} = \mathrm{D}_{s_+(x)}$ as well as $s_+(x) \in S_+$. Thus, assuming that the assertion is already proved for $x \in S_+$, the strong Markov property yields

$$\mathbf{P}_x\left(\left\{\int_0^{\mathrm{D}_{S_+}+t} \mathbf{1}_{S_+}(X_s)\mathrm{d}V_s(X) = d_+\left(x, \int_0^{\mathrm{D}_{S_+}+t} \mathbf{1}_{S_+}(X_s)\mathrm{d}s\right); \mathrm{D}_{S_+} < \infty\right\}\right)$$

$$= \mathbf{P}_x(\{\mathrm{D}_{S_+} < \infty\})\mathbf{P}_{s_+(x)}\left(\left\{\int_0^t \mathbf{1}_{S_+}(X_s)\mathrm{d}V_s(X) = d_+\left(x, \int_0^t \mathbf{1}_{S_+}(X_s)\mathrm{d}s\right)\right\}\right)$$

$$= \mathbf{P}_x(\{\mathrm{D}_{S_+} < \infty\}),$$

i.e. (4.11) holds.

So far we have proved, that it is indeed sufficient to prove the assertion of the Lemma for the case $x \in S_+$. Let $x \in S_+$ be fixed and put $c = \inf\{y \geq x : y \in K_-\}$. Then $c > x$ and we have $X \in [x,c]$ $\mathbf{P}_x$-a.s. In the case of $c < \infty$, $c \in E$ or $c \in K_- \setminus E$ holds. If $c \in K_- \setminus E$ then (2.5) implies that there exists some $\epsilon > 0$ such that $(c-\epsilon, c) \subseteq R$. Consequently, there is some n such that

$x < a_n < b_n = c$, $b_n \in K_- \setminus E$, $a_n \in K_+ \setminus E$. From this we obtain that $\mathbf{P}_x$-a.s.

$$\int_0^t \mathbf{1}_{S_+}(X_s)\mathrm{d}V_s(X) = \int_0^{t\wedge \mathrm{D}_c} \mathbf{1}_{S_+}(X_s)\mathrm{d}V_s(X)$$
$$\int_0^t \mathbf{1}_{S_+}(X_s)\mathrm{d}s = \int_0^{t\wedge \mathrm{D}_c} \mathbf{1}_{S_+}(X_s)\mathrm{d}s,\ t \geq 0.$$

Obviously, these equalities remain valid if we replace c by $d = c \wedge \sup\{y \geq x : \mathbf{P}_x(\{\mathrm{D}_y < \infty\}) = 1\}$ (see (2.9)). Passing, if necessary, to the stopped process X^{D_d}, without restricting the generality we can assume that the state interval I satisfies

(4.12) *$I = [x, d]$ with $E = \{d\}$ or $I = [x, d)$ (in the case that d does not belong to the original state interval), $\mathbf{P}_x(\{\mathrm{D}_y < \infty\}) = 1$ for every $y < d$.*

In particular, this yields $[x, d) \cap K_- = \emptyset$ and

$$[x, d) = S_+ \cup \bigcup_n [a_n, b_n),\ a_n,\ b_n \in K_+,\ R = \bigcup_n (a_n, b_n).$$

Denote

$$V_t^+ = \int_0^t \mathbf{1}_{S_+}(X_s)\mathrm{d}V_s(X), \qquad t \geq 0.$$

Then

(4.13) $$V_t^+ = \ell([X_0, X_t] \cap S_+),\ t \geq 0, a.s.$$

Indeed, in view of $L_-^X(\cdot, y) = 0$ for $y \in K_+$ (cf. (1.17)(iv)), (1.15) implies $\int_0^t \mathbf{1}_{S_+}(X_s)\mathrm{d}\langle M(X)\rangle_s = 0$ and thus $\int_0^t \mathbf{1}_{S_+}(X_s)\mathrm{d}M(X)_s = 0$, $t \geq 0$, $\mathbf{P}_x$-a.s. Taking into account (4.12) and (2.3)(iii) from this we obtain

$$\begin{aligned} V_t^+ &= \int_0^t \mathbf{1}_{S_+}(X_s)\mathrm{d}X_s = X_t - X_0 - \int_0^t \mathbf{1}_{\bigcup_n [a_n, b_n)}(X_s)\mathrm{d}X_s \\ &= X_t - X_0 - \int_0^t \sum_n \mathbf{1}_{[\mathrm{D}_{a_n}, \mathrm{D}_{b_n})}(s)\,\mathrm{d}X_s \\ &= X_t - X_0 - \ell\Big([X_0, X_t] \cap \bigcup_n [a_n, b_n)\Big) \\ &= \ell([X_0, X_t] \cap S_+), \end{aligned}$$

$t \geq 0$, $\mathbf{P}_x$-a.s.

Now let $L : [x, d) \to [0, \infty)$ be a continuous strictly increasing function satisfying $\sum_n (L(b_n) - L(a_n)) = \infty$ if $\sup_n b_n = d$. We put

$$\begin{aligned} A_t &= \int_{\left[X_0, \max_{s\leq t} X_s\right] \cap \bigcup_n [a_n, b_n)} L(\mathrm{d}y) + \int_0^t \mathbf{1}_{S_+}(X_s)\mathrm{d}s, \\ T_t &= \inf\{s \geq 0 : A_s > t\},\ t \geq 0. \end{aligned}$$

Then we can easily verify that A is a weakly additive functional of $(X, \mathbb{F})$ (cf. (3.4)) satisfying the conditions of Theorem (3.6). In particular, (3.6)(iii) is fulfilled. In fact, we have the relation

$$A_\infty = \int_{[X_0,d]\cap\bigcup_n [a_n,b_n)} L(dy) + \int_0^\infty 1_{S_+}(X_s)ds,$$

and for $\sup_n b_n = d$ we obtain $A_\infty = \infty$ from the required property of L. On the other hand, if $v = \sup_n b_n < d$ then $[v, d) \subseteq S_+$. For $D_d(X) = \infty$ this means $A_\infty = \infty$ because of $D_{v+} < \infty$ a.s. and $A_{D_{v+}+t} = A_{D_{v+}} + t$, whereas in the case of $D_d(X) < \infty$ it holds that $\lim_{t\to\infty} X_t = d \in E$.

Now, by (3.6) $Y = X \circ T$ turns out to be a continuous strong Markov process on $(\Omega, \mathcal{F}, \mathbf{P}_y, y \in I)$ with $Y_0 = X_0$ a.s. From the definition of A it follows that every point $y \in I$ is right singular for Y. Consequently, the process Y admits increasing paths and by (4.1) there exists a function $h : I \times \mathbb{R}_+ \to I$ such that

$$Y_t = h(Y_0, t), \qquad t \geq 0, \ a.s.$$

Recall that $h(y, \cdot)$ is continuous and strictly increasing on $[0, D_d(h(y, \cdot)))$ for every $y \in I$.

Now (4.13) implies

(4.14) $$V^+_{T_t} = \ell([x, h(x,t)] \cap S_+), \ t \geq 0, \ \mathbf{P}_x - a.s.$$

Now let us show that there exists a function $d_+(x, \cdot) : \mathbb{R}_+ \to [0, \infty]$ such that

(4.15) $$V^+_{T_t} = d_+\left(x, \int_0^t 1_{S_+}(Y_s)ds\right), \ t \geq 0, \ \mathbf{P}_x - a.s.$$

Let $g(x, \cdot) : [x, h(x, \infty)) \to [0, \infty)$ denote the right-inverse of the strictly increasing continuous function $h(x,t)$, $t \in [0, D_d(h(x, \cdot)))$. We put

$$g^*(x,t) := \int_0^t 1_{S_+}(s)\, g(x, ds), \qquad x \leq t < h(x, \infty).$$

Then $\mathbf{P}_x$-a.s. $g^*(x,t) = \int_0^{g(x,t)} 1_{S_+}(Y_s)ds$ and thus

(4.16) $$g^*(x, h(x,t)) = \int_0^t 1_{S_+}(Y_s)ds, \ t \geq 0, \mathbf{P}_x - a.s.$$

Now let $h^*(x, \cdot) : [0, \int_0^\infty 1_{S_+}(h(x,s))ds) \to [x, h(x, \infty))$ denote the right-inverse of the continuous increasing function $g^*(x, \cdot)$. Then

$$h^*(x, g^*(x,t)) = \begin{cases} t & : \ t \in S_+ \\ b_n & : \ t \in [a_n, b_n) \end{cases}, \ x \leq t < h(x, \infty),$$

and, consequently,

$$\ell\Big([x,h^*(x,g^*(x,t))]\cap S_+\Big)=\ell([x,t]\cap S_+),\qquad x\le t<h(x,\infty).$$

Using (4.14) and (4.16) from this we obtain

$$\begin{aligned} V^+_{T_t} &= \ell([x,h(x,t)]\cap S_+)\\ &= \ell\Big([x,h^*(x,g^*(x,h(x,t)))]\cap S_+\Big)\\ &= \ell\Big([x,h^*(x,\int_0^t \mathbf{1}_{S_+}(Y_s)\mathrm{d}s)]\cap S_+\Big),\ t\ge 0,\ \mathbf{P}_x-a.s. \end{aligned}$$

Defining

$$d_+(x,t):=\begin{cases} \ell([x,h^*(x,t)]\cap S_+)\ :\ 0\le t<\int_0^\infty \mathbf{1}_{S_+}(h(x,s))\mathrm{d}s,\\ \lim\limits_{v\uparrow\int_0^\infty \mathbf{1}_{S_+}(h(x,s))\mathrm{d}s} \ell([x,h^*(x,v)]\cap S_+)\ :\ \infty>t\ge\int_0^\infty \mathbf{1}_{S_+}(h(x,s))\mathrm{d}s, \end{cases}$$

we have just verified equation (4.15). It is easy to show that $d_+(x,\cdot)$ is continuous.

It was our goal to get an explicit expression for the process V^+. To this end, we apply the $\mathbb{F}\circ T$-time change A, which is the right-inverse of T, to (4.15),

$$V^+_{T_{A_t}}=d_+(x,\int_0^{A_t}\mathbf{1}_{S_+}(Y_s)\mathrm{d}s)=d_+(x,\int_0^t\mathbf{1}_{S_+}(X_{T_{A_s}})\mathrm{d}A_s),\qquad t\ge 0,$$

$\mathbf{P}_x$-a.s. We have $T_{A_s}(\omega)=s$ if and only if $(s,\omega)\in H_+(A)$ (cf. (3.3)) and thus,

$$\begin{aligned} &\int_0^t \mathbf{1}_{S_+}(X_{T_{A_s}})\mathrm{d}A_s\\ &= \int_0^t \mathbf{1}_{S_+}(X_s)\mathbf{1}_{H_+(A)}((s,\cdot))\mathrm{d}A_s+\int_0^t \mathbf{1}_{S_+}(X_{T_{A_s}})\mathbf{1}_{H_+^c(A)}((s,\cdot))\mathrm{d}A_s\\ &= \int_0^t \mathbf{1}_{S_+}(X_s)\mathrm{d}s, \end{aligned}$$

where we have also used the identity $\int_0^t \mathbf{1}_{H_+^c(A)}((s,\cdot))\mathrm{d}A_s=0$. This yields

(4.17)
$$V^+_{T_{A_t}}=d_+\Big(x,\int_0^t \mathbf{1}_{S_+}(X_s)\mathrm{d}s\Big),\ t\ge 0,$$

$\mathbf{P}_x$-a.s., and, to finish the proof, it remains to be verified that $V^+_{T_{A_t}}=V^+_t,\ t\ge 0,\ \mathbf{P}_x$-a.s. To this end, it suffices to show that V^+ is constant on those intervals where A is constant, $\mathbf{P}_x$-a.s.

For $t_1 < t_2$ on $\{A_{t_1} = A_{t_2}\}$ we have $\int_{t_1}^{t_2} \mathbf{1}_{S_+}(X_s)\mathrm{d}s = 0$ and, therefore, $X_s \in \bigcup_n [a_n, b_n) \cup E$ for Lebesgue-almost all $s \in [t_1, t_2]$ on $\{A_{t_1} = A_{t_2}\}$. Moreover, on $\{A_{t_1} = A_{t_2}\}$

$$\int_{(\max_{s\le t_1} X_s, \max_{s\le t_2} X_s)\cap\bigcup_n [a_n,b_n)} L(\mathrm{d}y) = 0,$$

holds, and, since L is strictly increasing, it follows that

$$(\max_{s\le t_1} X_s, \max_{s\le t_2} X_s) \cap \bigcup_n [a_n, b_n) = \emptyset,$$

i.e. (compare also (2.5))

$$[\max_{s\le t_1} X_s, \max_{s\le t_2} X_s) \subseteq S_+ \cup E.$$

In the case of $\max_{s\le t_1} X_s < \max_{s\le t_2} X_s$ this contradicts the fact that $X_s \in \bigcup_n [a_n, b_n) \cup E$ for Lebesgue-almost all $s \in [t_1, t_2]$. For $\max_{s\le t_1} X_s = \max_{s\le t_2} X_s$ it follows that $X_s \in \bigcup_n [a_n, b_n) \cup E$ for every $s \in [t_1, t_2]$, $\mathbf{P}_x$-a.s. on $\{A_{t_1} = A_{t_2}\}$.

This yields $V_{t_1}^+ = V_{t_2}^+$ on $\{A_{t_1} = A_{t_2}\}$ $\mathbf{P}_x$-a.s. □

Proof of Theorem (4.3). First, we observe that

$$I \setminus (R \cup \{a_n : a_n \in K_+\} \cup \{b_n : b_n \in K_-\} \cup S_+ \cup S_-) \subseteq E.$$

Now $X = X^{D_E}$ (cf. (2.3)(iii)) implies $V(X) = V(X)^{D_E}$ and thus

$$\int_0^t \mathbf{1}_E(X_s)\mathrm{d}V_s(X) = 0, \qquad t \ge 0, \ a.s.$$

From this and (4.5), (4.8) and (4.9) we obtain

$$\begin{aligned} V_t(X) &= \int_0^t \mathbf{1}_R(X_s)\mathrm{d}V_s(X) + \int_0^t \mathbf{1}_{\{a_n : a_n \in K_+\}}(X_s)\mathrm{d}V_s(X) \\ &\quad + \int_0^t \mathbf{1}_{\{b_n : b_n \in K_-\}}(X_s)\mathrm{d}V_s(X) + \int_0^t \mathbf{1}_{S_+}(X_s)\mathrm{d}V_s(X) \\ &\quad + \int_0^t \mathbf{1}_{S_-}(X_s)\mathrm{d}V_s(X) \\ &= 1/2 \int_R L^X(t,y)\mu_g(\mathrm{d}y) + \sum_{n: a_n \in K_+} L^X(t, a_n) - \sum_{n: b_n \in K_-} L^X(t, b_n) \\ &\quad + d_+\Big(x, \int_0^t \mathbf{1}_{S_+}(X_s)\mathrm{d}s\Big) - d_-\Big(x, \int_0^t \mathbf{1}_{S_-}(X_s)\mathrm{d}s\Big), \end{aligned}$$

$t \geq 0$, $\mathbf{P}_x$-a.s. To finish the proof of Theorem (4.3), it remains to show that d_+ and d_- are $\mathfrak{B}^u(I) \otimes \mathfrak{B}(\mathbb{R}_+)$-measurable. We postpone the proof of this fact to (4.25). □

Our next goal is to answer the question whether the functions g, d_+ and d_- in the above identity for $V(X)$ are uniquely determined by the process X. To this end we need the following lemma.

(4.18) Lemma *Let* $(X, \mathbb{F})$ *on* $(\Omega, \mathcal{F}, \mathbf{P}_x, x \in I)$ *be a continuous strong Markov process. Then, for every* $x \in I$ *and every* $z \geq x$ *(resp.* $z \leq x$*)*

$$\int_0^{\mathrm{D}_z} \mathbf{1}_{S_+}(X_s)ds \quad \left(\textit{resp.} \int_0^{\mathrm{D}_z} \mathbf{1}_{S_-}(X_s)ds \right)$$

is deterministic $\mathbf{P}_x$*-a.s. on* $\{\mathrm{D}_{S_+} < \infty\}$ *(resp.* $\{\mathrm{D}_{S_-} < \infty\}$*). In other words, if* $\mathbf{P}_x(\{\mathrm{D}_{S_\pm} < \infty\}) > 0$ *then*

$$\int_0^{\mathrm{D}_z} \mathbf{1}_{S_\pm}(X_s)ds = \mathbf{E}_x \int_0^{\mathrm{D}_z} \mathbf{1}_{S_\pm}(X_s)ds \Big/ \mathbf{P}_x(\{\mathrm{D}_{S_\pm} < \infty\})$$

$\mathbf{P}_x$*-a.s. on* $\{\mathrm{D}_{S_\pm} < \infty\}$*. Moreover, if* $\mathbf{P}_x(\{\mathrm{D}_{S_\pm} < \infty\}) > 0$ *then also*

$$\int_0^{\infty} \mathbf{1}_{S_\pm}(X_s)ds = \mathbf{E}_x \int_0^{\infty} \mathbf{1}_{S_\pm}(X_s)ds \Big/ \mathbf{P}_x(\{\mathrm{D}_{S_\pm} < \infty\})$$

$\mathbf{P}_x$*-a.s. on* $\{\mathrm{D}_{S_\pm} < \infty\}$*.*

Proof. We restrict ourselves to the proof of the assertion concerning the occupation time in the set S_+. Let $x \in I$ be fixed and suppose $\mathbf{P}_x(\{\mathrm{D}_{S_+} < \infty\}) > 0$. From (2.12) we have $\mathrm{D}_{S_+} = \mathrm{D}_{s_+(x)}$ $\mathbf{P}_x$-a.s. on $\{\mathrm{D}_{S_+} < \infty\}$, where $s_+(x) = \inf\{y \geq x : y \in S_+\}$. Therefore, for $x \leq z \leq s_+(x)$

$$\int_0^{\mathrm{D}_z} \mathbf{1}_{S_+}(X_s)ds = 0, \qquad \mathbf{P}_x - a.s.$$

Now let $z > s_+(x)$. Then, using the strong Markov property, for $B \in \mathfrak{B}(\mathbb{R}_+)$

$$\begin{aligned} &\mathbf{P}_x\left(\left\{\int_0^{\mathrm{D}_z} \mathbf{1}_{S_+}(X_s)ds \in B\right\} \cap \{\mathrm{D}_{S_+} < \infty\}\right) \\ &= \mathbf{P}_{s_+(x)}\left(\left\{\int_0^{\mathrm{D}_z} \mathbf{1}_{S_+}(X_s)ds \in B\right\}\right)\mathbf{P}_x(\{\mathrm{D}_{S_+} < \infty\}). \end{aligned}$$

Thus, to prove the assertion, it suffices to show that

$$\mathbf{P}_{s_+(x)}\left(\left\{\int_0^{\mathrm{D}_z} \mathbf{1}_{S_+}(X_s)ds \in B\right\}\right) \in \{0, 1\}.$$

In other words, for every $x \in S_+$ and every $z > x$ we have to verify that

$$\text{(4.19)} \qquad \int_0^{D_z} \mathbf{1}_{S_+}(X_s)ds \text{ is deterministic } \mathbf{P}_x - a.s.$$

We adopt some ideas from the proof of Lemma (4.9) and make use of the processes (A_t), (T_t) and $Y_t = X_{T_t} = h(X_0, t)$ introduced there. Observe, that in the definition of the processes A, T and Y we did not employ the semimartingale property of X.

In view of (2.9), $x \in S_+$ yields $\mathbf{P}_x(\{D_z < \infty\}) \in \{0,1\}$. First let $\mathbf{P}_x(\{D_z < \infty\}) = 1$. Then $\mathbf{P}_x$-a.s.

$$\begin{aligned}
\int_0^{D_z} \mathbf{1}_{S_+}(X_s)ds &= \int_0^{D_z} \mathbf{1}_{S_+}(X_s)dA_s \\
&= \int_0^{A_{D_z}} \mathbf{1}_{S_+}(X_{T_s})ds \\
&= \int_0^{D_z(Y)} \mathbf{1}_{S_+}(Y_s)ds \\
&= \int_0^{g(x,z)} \mathbf{1}_{S_+}(h(x,s))ds,
\end{aligned}$$

where $g(x, \cdot)$ denotes the inverse of $h(x, \cdot)$. Obviously, the right-hand side of the last identity is deterministic, which proves the assertion for $z > x$ and $\mathbf{P}_x(\{D_z < \infty\}) = 1$.

Now let

$$y = \sup\{z \geq x : \ z \in I, \mathbf{P}_x(\{D_z < \infty\}) = 1\}.$$

Because of $x \in S_+$ we have $x \in K_+ \setminus E$ and thus $y > x$. Of course, $y \leq \sup I$ holds. Now, if $\mathbf{P}_x(\{D_y < \infty\}) = 1$ then, in view of $\mathbf{P}_x(\{D_{y+\epsilon} < \infty\}) = 0$, for every $\epsilon > 0$ we get $y \in K_-$ and $[x, y) \cap K_- = \emptyset$. In the case of $y \in E$ it follows immediately that $\int_0^{D_z} \mathbf{1}_{S_+}(X_s)ds = \int_0^{D_y} \mathbf{1}_{S_+}(X_s)ds$ $\mathbf{P}_x$-a.s. for $z \geq y$, and, thus (4.19) holds true. If otherwise $y \in K_- \setminus E$, then using (2.5) and taking into account that $[x, y) \cap K_- = \emptyset$, we see that y has to be a right boundary point of a component of R:

$$x < a_n < b_n = y, \ (a_n, b_n) \subseteq R.$$

There, we have $a_n \in K_+$ and thus $a_n \notin S_+$, which yields again

$$\int_0^{D_z} \mathbf{1}_{S_+}(X_s)ds = \int_0^{D_y} \mathbf{1}_{S_+}(X_s)ds$$

$\mathbf{P}_x$-a.s. for $z \geq y$. Summarizing, we have proved (4.19) in the case of $\mathbf{P}_x(\{D_y < \infty\}) = 1$.

The case $\mathbf{P}_x(\{D_y = \infty\}) = 1$ remains to be treated. Because of $\mathbf{P}_x(\{\lim_{v \uparrow y} D_v = D_y\}) = 1$, for $z \geq y$, $z \in I$, we obtain the identity

$$\begin{aligned}\int_0^{D_z} \mathbf{1}_{S_+}(X_s)ds &= \int_0^{D_y} \mathbf{1}_{S_+}(X_s)ds \\ &= \lim_{v\uparrow y} \int_0^{D_v} \mathbf{1}_{S_+}(X_s)ds \\ &= \lim_{v\uparrow y} \int_0^{g(x,v)} \mathbf{1}_{S_+}(h(x,s))ds\end{aligned}$$

$\mathbf{P}_x$-a.s. This finishes the proof of (4.19).

Finally, the second assertion follows from

$$\int_0^{\infty} \mathbf{1}_{S_+}(X_s)ds = \lim_{z\uparrow \sup I} \int_0^{D_z} \mathbf{1}_{S_+}(X_s)ds \qquad \mathbf{P}_x - a.s.$$

□

Using the preceding Lemma our next goal is to clarify how the functions d_+, d_- in (4.3), (4.10) can be explicitly related with the process X. As a consequence we obtain that these functions are, in a certain sense, uniquely determined by X.

(4.20) Proposition *Suppose that the conditions of Theorem (4.3) are satisfied. Let $d_\pm : I \times \mathbb{R}_+ \to [0,\infty]$ be functions such that $d_\pm(x,\cdot)$ is measurable for every $x \in I$ and the identities (4.10) hold. Denote*

$$\begin{aligned}f_+(x,z) &= \mathbf{E}_x \int_0^{D_z} \mathbf{1}_{S_+}(X_s)ds \Big/ \mathbf{P}_x(\{D_{S_+} < \infty\}),\ z \geq x,^2 \\ f_-(x,z) &= \mathbf{E}_x \int_0^{D_z} \mathbf{1}_{S_-}(X_s)ds \Big/ \mathbf{P}_x(\{D_{S_-} < \infty\}),\ z \leq x,\end{aligned}$$

and

$$\begin{aligned}h_+^*(x,u) &= \inf\{z \geq x : f_+(x,z) \geq u\},\ 0 \leq u \leq f_+(x,\infty), \\ h_-^*(x,u) &= \sup\{z \leq x : f_-(x,z) \geq u\},\ 0 \leq u \leq f_-(x,-\infty).\end{aligned}$$

Then we have

$$\begin{aligned}d_+(x,u) &= \ell([x,h_+^*(x,u)] \cap S_+),\ 0 \leq u \leq f_+(x,\infty), \\ d_-(x,u) &= \ell([h_-^*(x,u),x] \cap S_-),\ 0 \leq u \leq f_-(x,-\infty).\end{aligned}$$

Proof. We only show the assertion for d_+. First, let us verify that for every $x \in I$

---

[2] Observe (1.28).

(4.21) $f_+(x,z)$ is continuous in $z \geq x$.[3]

In fact, in the case of $\mathbf{P}_x(\{\mathrm{D}_{S_+} < \infty\}) = 0$ we have $f_+(x,z) = 0$, $z \geq x$. For $\mathbf{P}_x(\{\mathrm{D}_{S_+} < \infty\}) > 0$ and in view of (2.12)(i) we get $f_+(x,z) = 0$, $x \leq z \leq s_+(x)$, and, for $z \geq s_+(x)$ the strong Markov property yields the identity

$$f_+(x,z) = \mathbf{E}_{s_+(x)} \int_0^{\mathrm{D}_z} \mathbf{1}_{S_+}(X_s)\mathrm{d}s.$$

Therefore, (4.21) remains to be shown only for $x = s_+(x) \in S_+$. As in the proof of Lemma (4.18) let

(4.22) $$y = \sup\{z \geq x : z \in I,\ \mathbf{P}_x(\{\mathrm{D}_z < \infty\}) = 1\}$$

(cf. also (2.9)). In view of $x \in S_+$ we have $y > x$. For $x \leq z < y$, $z \notin K_-$ and $\mathbf{E}_x \mathrm{D}_z < \infty$ hold (cf. (2.9)). In particular, this implies $\mathbf{P}_x(\{\mathrm{D}_z = \mathrm{D}_{z+} = \lim_{\epsilon\downarrow 0} \mathrm{D}_{z+\epsilon} < \infty\}) = 1$, thus, for $x \leq z < y$ the continuity of $f_+(x,z)$ is obvious. Now, if moreover $\mathbf{P}_x(\{\mathrm{D}_y < \infty\}) = 1$, then from the definition of y we get $y \in K_-$, consequently

(4.23) $$f_+(x,z) = f_+(x, z \wedge y),\ z \geq x,$$

which yields (4.21). If otherwise (cf. (2.9)) $\mathbf{P}_x(\{\mathrm{D}_y < \infty\}) = 0$, then (4.23) is immediate, and again (4.21) holds true.

To prove (4.20), we first assume $x \in S_+$. In the proof of Lemma (4.9) we have shown that

$$\int_0^t \mathbf{1}_{S_+}(X_s)\mathrm{d}V_s(X) = \ell([x, X_t] \cap S_+), \qquad t \geq 0,\ \mathbf{P}_x - a.s.$$

(cf. (4.13)). By (2.9), $\mathbf{P}_x(\{\mathrm{D}_z < \infty\}) \in \{0,1\}$ for every $z \geq x$. For y as defined in (4.22) we then get

$$\ell([x, X_{\mathrm{D}_z}] \cap S_+) = \ell([x, z \wedge y] \cap S_+), \qquad z \geq x,\ \mathbf{P}_x - a.s.$$

This implies $\mathbf{P}_x$-a.s.

$$d_+\Big(x, \int_0^{\mathrm{D}_z} \mathbf{1}_{S_+}(X_s)\mathrm{d}s\Big) = \int_0^{\mathrm{D}_z} \mathbf{1}_{S_+}(X_s)\mathrm{d}V_s(X) = \ell([x, z\wedge y] \cap S_+),\ z \geq x.$$

Using (4.18), for the point $x \in S_+$ we have $\mathbf{P}_x$-a.s.

$$\int_0^{\mathrm{D}_z} \mathbf{1}_{S_+}(X_s)\mathrm{d}s = \mathbf{E}_x \int_0^{\mathrm{D}_z} \mathbf{1}_{S_+}(X_s)\mathrm{d}s = f_+(x,z), \qquad z \geq x,$$

and, consequently,

$$d_+(x, f_+(x,z)) = \ell([x, z\wedge y] \cap S_+), \qquad z \geq x.$$

[3] In the topology of $\mathbb{R} \cup \{\infty\}$.

The continuity of $f_+(x,\cdot)$ and (4.23) imply

$$\begin{aligned} f_+(x, h_+^*(x,u)) &= u \\ h_+^*(x,u) &\le y, \end{aligned}$$

for every $0 \le u \le f_+(x,\infty)$. Now this yields

$$d_+(x,u) = \ell([x, h_+^*(x,u)] \cap S_+).$$

Finally, consider the case that $x \in I$ is arbitrary. In the case of $\mathbf{P}_x(\{\mathrm{D}_{S_+} < \infty\}) = 0$ there is nothing to show. So let $\mathbf{P}_x(\{\mathrm{D}_{S_+} < \infty\}) > 0$. Recall, that in view of (2.12), $\mathrm{D}_{S_+} = \mathrm{D}_{s_+(x)} \mathbf{P}_x$-a.s. on $\{\mathrm{D}_{S_+} < \infty\}$, where $s_+(x) = \inf\{z \ge x : z \in S_+\}$. Then we have $f_+(x,z) = 0$ if $x \le z \le s_+(x)$, and, for $z > s_+(x)$ applying the strong Markov property we obtain $f_+(x,z) = f_+(s_+(x),z)$. Altogether we have

$$f_+(x,z) = f_+(s_+(x), z \vee s_+(x)), \qquad z \ge x,$$

and from this

$$\textbf{(4.24)} \qquad \begin{aligned} f_+(x,\infty) &= f_+(s_+(x),\infty) \\ s_+(x) \vee h_+^*(x,u) &= h_+^*(s_+(x),u), \ 0 \le u \le f_+(x,\infty). \end{aligned}$$

Using again (4.18) and the strong Markov property, for $z \ge s_+(x)$ we obtain $\mathbf{P}_x$-a.s.

$$\begin{aligned} & d_+(x, f_+(x,z)) \mathbf{1}_{\{\mathrm{D}_{S_+}<\infty\}} \\ &= d_+\left(x, \int_0^{\mathrm{D}_z} \mathbf{1}_{S_+}(X_s) ds\right) \mathbf{1}_{\{\mathrm{D}_{S_+}<\infty\}} \\ &= \left[\mathbf{E}_x d_+\left(x, \int_0^{\mathrm{D}_z} \mathbf{1}_{S_+}(X_s) ds\right) \Big/ \mathbf{P}_x(\{\mathrm{D}_{S_+} < \infty\})\right] \mathbf{1}_{\{\mathrm{D}_{S_+}<\infty\}} \\ &= \left[\mathbf{E}_x \int_0^{\mathrm{D}_z} \mathbf{1}_{S_+}(X_s) dV_s(X) \Big/ \mathbf{P}_x(\{\mathrm{D}_{S_+} < \infty\})\right] \mathbf{1}_{\{\mathrm{D}_{S_+}<\infty\}} \\ &= \left[\mathbf{E}_{s_+(x)} \int_0^{\mathrm{D}_z} \mathbf{1}_{S_+}(X_s) dV_s(X)\right] \mathbf{1}_{\{\mathrm{D}_{S_+}<\infty\}} \\ &= \left[\mathbf{E}_{s_+(x)} d_+\left(s_+(x), \int_0^{\mathrm{D}_z} \mathbf{1}_{S_+}(X_s) ds\right)\right] \mathbf{1}_{\{\mathrm{D}_{S_+}<\infty\}} \\ &= d_+\Big(s_+(x), f_+(s_+(x), z)\Big) \mathbf{1}_{\{\mathrm{D}_{S_+}<\infty\}}, \end{aligned}$$

i.e.

$$d_+\Big(x, f_+(x,z)\Big) = d_+\Big(s_+(x), f_+\big(s_+(x), z \vee s_+(x)\big)\Big).$$

Taking into account (4.21) and the fact that the assertion is already verified for $s_+(x) \in S_+$, for $0 \le u \le f_+(x,\infty)$ we obtain

$$\begin{aligned} d_+(x,u) &= d_+(s_+(x),u) = \ell\Big(\big[s_+(x), h_+^*(s_+(x),u)\big] \cap S_+\Big) \\ &= \ell\big([x, h_+^*(x,u)] \cap S_+\big), \end{aligned}$$

where we have also used (4.24) and the relation $[x, s_+(x)) \cap S_+ = \emptyset$. □

(4.25) Corollary *Suppose that the conditions of Theorem (4.3) are satisfied. Let $d_\pm|I \times \mathbb{R}_+ \to [0,\infty]$ be functions such that $d_\pm(x,\cdot)$ is measurable for every $x \in I$ and the identities (4.10) are fulfilled. Then the following assertions hold.*

(i) *For every $x \in I$ the function $d_+(x,\cdot)$ restricted to $[0, f_+(x,\infty)] = \big[0, \mathbf{E}_x \int_0^\infty \mathbf{1}_{S_+}(X_s)ds / \mathbf{P}_x(\{\mathrm{D}_{S_+} < \infty\})\big]$ is uniquely determined by the process X. The same holds true for $d_-(x,\cdot)$ restricted to $[0, f_-(x,-\infty)] = \big[0, \mathbf{E}_x \int_0^\infty \mathbf{1}_{S_-}(X_s)ds / \mathbf{P}_x(\{\mathrm{D}_{S_-} < \infty\})\big]$.*

(ii) *Without restricting the generality, we can assume that*

$$d_\pm(x,u) = d_\pm\big(x, u \wedge f_\pm(x, \pm\infty)\big), \; u \ge 0. \tag{4.26}$$

Then, the functions $d_\pm|I \times \mathbb{R}_+ \to [0,\infty]$ are $\mathfrak{B}^u(I) \otimes \mathfrak{B}(\mathbb{R}_+)$-measurable and continuous in the second variable.

The proof of this Corollary is immediate from Proposition (4.20).

The assertion of Corollary (4.25)(i) now justifies the following definition.

(4.27) Definition We call the functions $d_+|I \times \mathbb{R}_+ \to [0,\infty]$ resp. $d_-|I \times \mathbb{R}_+ \to [0,\infty]$ satisfying (4.10) and (4.26) *S_+-drift function* resp. *S_--drift function* of X.

(4.28) Remark The function g in Theorem (4.3) (cf. (4.4)) is not uniquely determined. Indeed, since on every component of R the scale function p of X is only determined up to an affine transformation, we have that g is determined only up to multiplication by a constant on every component of R. However, this implies that the set function μ_g generated from g (cf. (A2.6)) is independent from the particular choice of the scale of X.

In this section we completely discovered the intrinsic structure of the process $V(X)$ appearing in the semimartingale decomposition of X. Such a complete characterization of the local martingale part $M(X)$ seems to be impossible. In a certain sense we can analyse the process $M(X)$ using the so-called occupation time formula of Chapter V. We will return to this topic in Chapter VII.

However, closing this chapter, we give an elementary result on the structure of the local martingale part $M(X)$.

(4.29) Proposition *We have*

$$M(X)_t = \int_0^t \mathbf{1}_R(X_s)\mathrm{d}M(X)_s, \qquad t \geq 0, \ a.s.$$

Proof. It suffices to show that

$$\int_0^t \mathbf{1}_{I\setminus R}(X_s)\mathrm{d}M(X)_s = 0, \qquad t \geq 0, \ a.s.$$

Using (1.15) we obtain

$$\begin{aligned}
&\int_0^t \mathbf{1}_{I\setminus R}(X_s)\mathrm{d}\langle M(X)\rangle_s \\
&= \int_0^t \mathbf{1}_{K_+}(X_s)\mathrm{d}\langle M(X)\rangle_s + \int_0^t \mathbf{1}_{K_-\setminus E}(X_s)\mathrm{d}\langle M(X)\rangle_s \\
&= \int_{K_+} L_-^X(t,y)\mathrm{d}y + \int_{K_-\setminus E} L_+^X(t,y)\mathrm{d}y, \quad t \geq 0, a.s.
\end{aligned}$$

But in view of (1.17)(iv), for every $y \in K_+$ (resp $y \in K_-$) we have $L_-^X(\cdot,y) = 0$ (resp. $L_+^X(\cdot,y) = 0$) which yields the assertion. □

V. Occupation Time Formula

V.1 Speed Measure

Let $(X, \mathbb{F})$ on $(\Omega, \mathcal{F}, \mathbf{P}_x, x \in I)$ be a continuous strong Markov process. We do *not* suppose $X \in \mathcal{S}(\mathbb{F}^X)$. The present section is devoted to the so-called speed measure of X which is basic for the behaviour in "time" of the process X, and, in particular for the computation of the occupation time of the process X in certain sets of the state space.

The notion of the speed measure is a well-known concept for regular continuous strong Markov processes (cf. [28],[31]). In this section we introduce a suitable speed measure without any restrictions on the underlying continuous strong Markov process that will be most helpful in what follows.

Our speed measure will be a measure on $(I \setminus E, \mathfrak{B}(I \setminus E))$. Its definition (see (5.13)) is closely related to subsets of $I \setminus E$ coming from the classification of points introduced in Chapter II. The definition requires some preparatory results which we prove at the beginning.

(5.1) Proposition *Let p be a scale function of X.*

(i) *There exists a function $f : I \to \mathbb{R}$ such that $(f(X_t) + t)_{t \geq 0} \in \mathcal{M}(\mathbb{F}^X, \mathrm{D}_{R^c})$. Moreover, f is continuous and strictly p-concave on each component (a, b) of R, that is*

$$f(x) > f(z)\frac{p(y) - p(x)}{p(y) - p(z)} + f(y)\frac{p(x) - p(z)}{p(y) - p(z)}$$

for all $a < z < x < y < b$.

(ii) *Let (a, b) denote an arbitrary component of R. If $a \in K_+ \setminus E$ then there exists a continuous function $f_a : [a, b) \to \mathbb{R}$ such that $((f_a(X_t) + t)_{t \geq 0}, \mathbb{F}^X)$ is a continuous local $\mathbf{P}_x$-martingale up to D_b for all $x \in [a, b)$. By symmetry, if $b \in K_- \setminus E$ then the above is true under the assumption that we replace f_a, $[a, b)$ and D_b by f_b, $(a, b]$ and D_a, respectively. Here f_a (resp. f_b) is strictly decreasing (resp. increasing) as well as strictly p-concave on (a, b). Finally, if a is attracting for (a, b) then f_a is even strictly p-concave on the extended component $[a, b)$ and the corresponding assertion holds if b is attracting for (a, b).*

Proof. (i) For $x \notin R$ we set $f(x) = 0$ and so f remains to be defined on R. We give the construction of the desired f on an arbitrary but fixed component (a, b) of R.

Choose $a_n \downarrow a$, $a_n > a$, $b_n \uparrow b$, $b_n < b$, set $S_n = \mathrm{D}_{a_n-} \wedge \mathrm{D}_{b_n+}$ and define

$$f_n(x) = \mathbf{E}_x S_n, \qquad x \in [a_n, b_n].$$

In view of (2.19)(ii), $f_n(x) < \infty$ holds and (2.21)(ii) yields the continuity of f_n on (a_n, b_n). Now, for $x \in (a_n, b_n) \subseteq (a_m, b_m)$ the strong Markov property as well as (2.17) imply

$$\begin{aligned} f_m(x) - f_n(x) &= \mathbf{E}_x(S_m - S_n) = \mathbf{E}_x(S_m \circ \theta_{S_n}) = \mathbf{E}_x \mathbf{E}_{X_{S_n}} S_m \\ &= f_m(a_n)\frac{p(b_n) - p(x)}{p(b_n) - p(a_n)} + f_m(b_n)\frac{p(x) - p(a_n)}{p(b_n) - p(a_n)}. \end{aligned}$$

Hence, there exist sequences $(c_n), (d_n)$ such that

$$f_n(x) + c_n p(x) + d_n = f_m(x) + c_m p(x) + d_m, \quad x \in (a_n, b_n) \subseteq (a_m, b_m).$$

We define the function f on (a, b) by

$$f(x) = f_n(x) + c_n p(x) + d_n \quad \text{if} \quad x \in (a_n, b_n).$$

Obviously, f is continuous. To show that f is strictly p-concave we only have to establish that f_n is strictly p-concave on (a_n, b_n). But, for $a_n < z < x < y < b_n$, applying the strong Markov property and (2.17) again, we obtain

$$\begin{aligned} f_n(x) &= \mathbf{E}_x \mathrm{D}_z \wedge \mathrm{D}_y + \mathbf{E}_z S_n \mathbf{P}_x(\{\mathrm{D}_z < \mathrm{D}_y\}) + \mathbf{E}_y S_n \mathbf{P}_x(\{\mathrm{D}_y < \mathrm{D}_z\}) \\ &> f_n(z)\frac{p(y) - p(x)}{p(y) - p(z)} + f_n(y)\frac{p(x) - p(z)}{p(y) - p(z)}. \end{aligned}$$

It remains to show that $(f(X_t) + t)_{t \geq 0} \in \mathcal{M}(\mathbb{F}^X, \mathrm{D}_{R^c})$. If $x \in R^c$ then with respect to $\mathbf{P}_x$ there is nothing to prove. We fix $x \in (a, b)$ for a certain component (a, b) of R. In order to verify that $((f(X_t) + t)_{t \geq 0}, \mathbb{F}^X)$ is a continuous local $\mathbf{P}_x$-martingale up to $\mathrm{D}_a \wedge \mathrm{D}_b$ we take as localizing sequence the stopping times S_n introduced in the definition of f on (a, b). Then, with $f_n(x) = \mathbf{E}_x S_n$,

$$\begin{aligned} \mathbf{E}_x(S_n | \mathcal{F}_t^X) &= \mathbf{E}_x(S_n | \mathcal{F}_t^X)\mathbf{1}_{\{t < S_n\}} + S_n \mathbf{1}_{\{t \geq S_n\}} \\ &= (\mathbf{E}_{X_t} S_n + t)\mathbf{1}_{\{t < S_n\}} + S_n \mathbf{1}_{\{t \geq S_n\}} \\ &= f_n(X_{t \wedge S_n}) + (t \wedge S_n), \end{aligned}$$

where we have used that $f_n(X_{S_n}) = 0$ $\mathbf{P}_x$-a.s. This means that $((f_n(X_{t \wedge S_n}) + (t \wedge S_n))_{t \geq 0}, \mathbb{F}^X)$ is a uniformly integrable $\mathbf{P}_x$-martingale. Due to (2.20)(i) the same is true for $((p(X_{t \wedge S_n}))_{t \geq 0}, \mathbb{F}^X)$ and reminding of the definition of f the assertion follows.

(ii) By symmetry, it suffices to consider the case $a \in K_+ \setminus E$. We choose $b_n \in (a, b)$ with $b_n \uparrow b$ and set $S_n = \mathrm{D}_{b_n+}$. Now define $f_n(x) = \mathbf{E}_x S_n$, $x \in [a, b)$. Then (2.22) yields that f_n is finite on $[a, b)$ while (2.21)(ii) ensures the continuity of f_n on (a, b). But f_n is even continuous at the point a. Indeed, for $x_k \downarrow a$ from $\mathbf{P}_a(\{\lim \mathrm{D}_{x_k} = 0\}) = 1$ and the strong Markov property, as in the proof of (2.21)(ii) we can deduce that

$$f_n(a) = \lim_{k\to\infty} \mathbf{E}_a \mathbf{1}_{\{\mathrm{D}_{x_k} < S_n\}} S_n = \lim_{k\to\infty} \mathbf{E}_{x_k} S_n = \lim_{k\to\infty} f_n(x_k).$$

Now define the function $f_a : [a, b) \to \mathbb{R}$ by

$$f_a(x) = f_n(x) - \mathbf{E}_{b_1} S_n \quad \text{if} \quad x \in [a, b_n).$$

This definition is correct because, for $m > n$, the strong Markov property gives us

$$f_m(x) - \mathbf{E}_{b_1} S_m = f_n(x) - \mathbf{E}_{b_1} S_n$$

for all $x \in [a, b_n) \subseteq [a, b_m)$. Now the rest of the proof is analogous to part (i) after the definition of f. We only add two comments. First, f_a is strictly decreasing since f_n is strictly decreasing on $[a, b_n]$. Second, if a is attracting for (a, b) then it follows easily from the continuity of f_a at a that f_a is strictly p-concave on $[a, b)$. □

The next lemma shows that the functions from (5.1) are in a certain sense uniquely determined.

(5.2) Lemma (i) *Let $f : I \to \mathbb{R}$ be a function such that $(f(X_t) + t)_{t\geq 0} \in \mathcal{M}(\mathbb{F}^X, \mathrm{D}_{R^c})$ and let f be continuous on R. Then f is unique on each component of R up to addition or subtraction of a scale function of X.*

(ii) *Let (a, b) denote an arbitrary component of R with $a \in K_+ \setminus E$. Let $f_a : [a, b) \to \mathbb{R}$ be a continuous function such that $((f_a(X_t) + t)_{t\geq 0}, \mathbb{F}^X)$ is a continuous local $\mathbf{P}_x$-martingale up to D_b for all $x \in [a, b)$. Then f_a is unique on $[a, b)$ up to an additive constant. An analogous result is true for $b \in K_- \setminus E$.*

Proof. (i) Fix two functions f_1, f_2 satisfying the required conditions. Then, $(f_1 - f_2)(X) \in \mathcal{M}(\mathbb{F}^X, \mathrm{D}_{R^c})$ holds and $f_1 - f_2$ is locally bounded on R. Now, the assertion follows from (2.20)(ii).

(ii) We restrict ourselves to the case $a \in K_+ \setminus E$. Again we fix two functions f_a^1, f_a^2 satisfying the required conditions and set

$$f(x) = \begin{cases} (f_a^1 - f_a^2)(x) & : \; x \in [a, b), \\ 0 & : \; \text{otherwise.} \end{cases}$$

Then $f(X) \in \mathcal{M}(\mathbb{F}^X, \mathrm{D}_{R^c})$, f is locally bounded on R and applying (2.20)(ii) once again, we obtain

$$f_a^1(x) = f_a^2(x) + c\,p(x) + d, \qquad x \in (a, b).$$

Setting $p(a) = \lim_{x\downarrow a} p(x)$ the continuity of f_a^1, f_a^2 on $[a, b)$ leads to

$$f_a^1 = f_a^2 + cp + d \quad \text{on} \quad [a, b).$$

If a is nonattracting, i.e. $p(a) = -\infty$, then c must vanish and we are done. Otherwise, if a is attracting then $cp(X)+d$ is a continuous local $\mathbf{P}_x$-martingale up to D_b for all $x \in [a, b)$. This implies

$$cp(a) + d = \mathbf{E}_a(cp(X_0) + d) = \mathbf{E}_a(cp(X_{\mathrm{D}_y}) + d) = cp(y) + d$$

for $y \in (a, b)$ proving $c = 0$ in that case, too. □

(5.3) Lemma *Let p denote a scale function of X.*

(i) *Consider a function $f : I \to \mathbb{R}$ satisfying the assertion of (5.1)(i). Then for all $x \in R$ there exists the right-hand derivative $D_p^+ f$ of f w.r.t. p:*

$$D_p^+ f(x) = \lim_{h \downarrow 0} \frac{f(x+h) - f(x)}{p(x+h) - p(x)},$$

and $D_p^+ f$ is strictly decreasing on each component of R.

(ii) *Let (a, b) denote a component of R, $a \in K_+ \setminus E$ and let f_a be a function satisfying the assertion of (5.1)(ii). We set $p(a) = \lim_{x \downarrow a} p(x)$. Then there exists the right-hand derivative of f_a w.r.t. p at the point a:*

$$D_p^+ f_a(a) = \lim_{h \downarrow 0} \frac{f_a(a+h) - f_a(a)}{p(a+h) - p(a)}.$$

Analogously, if $b \in K_- \setminus E$ then for f_b satisfying (5.1)(ii) there exists the left-hand derivative $D_p^- f_b$ of f_b w.r.t. p at the point b:

$$D_p^- f_b(b) = \lim_{h \downarrow 0} \frac{f_b(b-h) - f_b(b)}{p(b-h) - p(b)},$$

where $p(b) = \lim_{x \uparrow b} p(x)$. Moreover, we have $D_p^+ f_a(a) \le 0$ and $D_p^- f_b(b) \ge 0$.

(iii) *Let (a, b) denote a component of R. Fix functions f^1, f^2 satisfying 5.1(i) and functions f_a^1, f_a^2 and f_b^1, f_b^2 as in 5.1(ii). Further let p_1, p_2 denote two scale functions of X. Then it holds that*

$$D_{p_1}^+ f^1 = c D_{p_2}^+ f^2 + d, \; D_{p_1}^+ f^{1} = D_{p_1}^+ f^2 + d \quad \text{on} \quad (a, b),$$

$$D_{p_1}^+ f_a^1(a) = c D_{p_2}^+ f_a^2(a), \; D_{p_1}^+ f_a^1(a) = D_{p_1}^+ f_a^2(a),$$

and

$$D_{p_1}^+ f_b^1(b) = c D_{p_2}^+ f_b^2(b), \; D_{p_1}^+ f_b^1(b) = D_{p_1}^+ f_b^2(b)$$

where $c > 0$ and d are constants.

Proof. (i) For $x \in R$ the quotient

$$\frac{f(x+h)-f(x)}{p(x+h)-p(x)}$$

increases as $h \downarrow 0$ since f is p-concave. Therefore, for $h \downarrow 0$ the limit $D_p^+ f(x)$ exists. Since f is p-concave the quotient above is decreasing in x on every component of R. The fact that f is even strictly p-concave yields that $D_p^+ f$ is strictly decreasing on every component of R.

(ii) We only consider the case $a \in K_+ \setminus E$. If a is nonattracting for (a,b) then obviously $D_p^+ f_a(a) = 0$. Otherwise, if a is attracting for (a,b) then the quotient

$$\frac{f_a(a+h)-f_a(a)}{p(a+h)-p(a)}$$

increases for $h \downarrow 0$. But f_a decreases on $[a,b)$ and, consequently, this quotient is bounded above by zero proving part (ii) of the lemma.

(iii) immediately follows from (5.2). □

If $R = \bigcup_n (a_n, b_n)$ is the decomposition of the set R of regular points into its components then we define

$$\textbf{(5.4)} \qquad \bar{R} = \bigcup_n (a_n, b_n) \cup \Big(\{a_n : a_n \in K_+\} \cup \{b_n : b_n \in K_-\} \Big).$$

On the measurable space $(\bar{R} \setminus E, \mathfrak{B}(\bar{R} \setminus E))$ we now introduce a measure which will turn out to be the restriction of the speed measure of X to $\bar{R} \setminus E$.

Let f denote a function as in (5.1)(i). For $a_n \in K_+ \setminus E$ and $b_n \in K_- \setminus E$ let f_{a_n} and f_{b_n} be functions as in (5.1)(ii). Furthermore, we fix a scale function p of X.

We define the measure m_r on $(\bar{R} \setminus E, \mathfrak{B}(\bar{R} \setminus E))$ by[1]

$$\textbf{(5.5)} \qquad \begin{aligned} m_r(A) &= -\frac{1}{2}\int_{A \cap R} \mathrm{d}D_p^+ f - \sum_{a_n \in K_+ \setminus E} D_p^+ f_{a_n}(a_n)\delta_{a_n}(A) \\ &\quad + \sum_{b_n \in K_- \setminus E} D_p^- f_{b_n}(b_n)\delta_{b_n}(A). \end{aligned}$$

Lemma (5.3) ensures that m_r is well-defined.

We list some elementary properties of the measure m_r:

(5.6) Proposition (i) *If the scale function p of X is fixed then the measure m_r is unique.*

(ii) $m_r(G) > 0$ *for every open subset $G \subseteq R$.*

(iii) $m_r(K) < \infty$ *for every compact interval $K \subseteq R$.*

[1] δ_x denotes the Dirac measure at the point x.

(iv) *Let* (a,b) *denote a component of* R. *If* $a \in K_+ \setminus E$ *(resp.* $b \in K_- \setminus E$*) then it holds that* $m_r([a, a+\varepsilon)) < \infty$ *(resp.* $m_r((b-\varepsilon, b]) < \infty$*) for* $\varepsilon > 0$ *sufficiently small.*

(v) *Consider another scale* $\widetilde{p}$ *and functions* $\widetilde{f}, \widetilde{f}_{a_n}, \widetilde{f}_{b_n}$ *satisfying (5.1)(i) and (5.1)(ii), respectively. Denote by* $\widetilde{m}_r$ *the corresponding measure defined by (5.5). Then on each component* (a,b) *of* R *the measures* $\widetilde{m}_r$ *and* m_r *distinguish from each other only by a factor* $c > 0$, *that is* $m_r|_{(a,b)} = c\widetilde{m}_r|_{(a,b)}$. *If* $a \in K_+ \setminus E$ *(resp.* $b \in K_- \setminus E$*) then this is even true for the restrictions of* $\widetilde{m}_r$ *and* m_r *to* $[a,b)$ *(resp.* $(a,b]$*).*

Proof. (i) follows immediately from (5.3)(iii).

(ii) is straightforward since $D_p^+ f$ is strictly decreasing on each component of R and the assertion (iii) is obvious.

(iv) By symmetry we only deal with the case $a \in K_+ \setminus E$. Let f and f_a be functions as in (5.1)(i) and (5.1)(ii), respectively. As in the proof of (5.2)(ii) we conclude that

$$f = f_a + cp + d \quad \text{on} \quad (a,b)$$

where c, d are constants. This gives $D_p^+ f = D_p^+ f_a + c$ on (a,b). Since $D_p^+ f$ decreases on (a,b) this implies

$$\lim_{x \downarrow a} D_p^+ f(x) = \lim_{x \downarrow a} D_p^+ f_a(x) + c = D_p^+ f_a(a) + c > -\infty$$

by (5.3)(ii). Now the definition of m_r yields

$$\begin{aligned} m_r([a, a+\varepsilon]) &= -\frac{1}{2}\left[D_p^+ f(a+\varepsilon) - \lim_{x \downarrow a} D_p^+ f(x)\right] - D_p^+ f_a(a) \\ &< \infty \end{aligned}$$

for $a + \varepsilon < b$.

Finally, (v) turns out to be an immediate consequence of (5.3)(iii). □

We proceed with the definition of the speed measure of X and introduce measures m_+^S and m_-^S on $(S_+, \mathfrak{B}(S_+))$ and $(S_-, \mathfrak{B}(S_-))$ which will correspond with the restrictions of the speed measure to S_+ and S_-, respectively. We will make use of (2.8) which states that $I \setminus E$ admits a decomposition

$$I \setminus E = \bigcup_n J_n$$

into disjoint intervals satisfying either $J_n = [a,b]$, $(a,b) \subseteq R$, $a \in K_+ \setminus E$, $b \in K_- \setminus E$ or $J_n \setminus R \subseteq K_+ \setminus E$ or $J_n \setminus R \subseteq K_- \setminus E$. Recall that all points of such an interval J_n are connected.

(5.7) Proposition *There exist unique σ-finite measures* m_+^S *and* m_-^S *on* $(S_+, \mathfrak{B}(S_+))$ *and* $(S_-, \mathfrak{B}(S_-))$*, respectively, such that for every* n *and all* $y, z \in J_n$ *with* $y < z$ *it holds that*[2]

$$m_+^S([y,z] \cap S_+) = \mathbf{E}_y \int_0^{\mathrm{D}_z} \mathbf{1}_{S_+}(X_s)\,\mathrm{d}s / \mathbf{P}_y(\{\mathrm{D}_{S_+} < \infty\})$$

and

$$m_-^S([y,z] \cap S_-) = \mathbf{E}_z \int_0^{\mathrm{D}_y} \mathbf{1}_{S_-}(X_s)\,\mathrm{d}s / \mathbf{P}_z(\{\mathrm{D}_{S_-} < \infty\}).$$

Moreover, we have $m_+^S(\{y\}) = 0$ *and* $m_-^S(\{y\}) = 0$ *for all* $y \in S_+$ *and* $y \in S_-$*, respectively.*

Proof. It suffices to show the existence of m_+^S since the proof for m_-^S is analogous.

First, we verify the following assertion:

(5.8) In the case of $J_n \setminus R \subseteq K_+ \setminus E$ for $x \in J_n \setminus R$ and all $z > x$, $z \in J_n$, it holds that $\mathbf{P}_x(\{\mathrm{D}_z < \infty\}) = 1$.

From (2.9) we know that $\mathbf{P}_x(\{\mathrm{D}_z < \infty\}) \in \{0,1\}$ and we are done if $y := \sup\{z \geq x : \mathbf{P}_x(\{\mathrm{D}_z < \infty\}) > 0\}$ is not an element of J_n. But, $y \in J_n$ leads to a contradiction. Indeed, if $y \in J_n$ then $y \in K_+ \setminus E$ since y cannot be regular. Now $\mathbf{P}_x(\{\mathrm{D}_y < \infty\}) = 0$ is impossible because x and y are connected. So only the case $y \in J_n$, $y \in K_+ \setminus E$, $\mathbf{P}_x(\{\mathrm{D}_y < \infty\}) > 0$ remains to be considered. It is an easy consequence of (2.3)(i) that if $y \in K_+ \setminus E$ then there exists a point $z > y$ with $\mathbf{P}_y(\{\mathrm{D}_z < \infty\}) > 0$. Using the strong Markov property we obtain $\mathbf{P}_x(\{\mathrm{D}_z < \infty\}) > 0$ which is a contradiction of the definition of y proving (5.8).

As in (4.20) set

$$f_+(x,y) = \mathbf{E}_x \int_0^{\mathrm{D}_y} \mathbf{1}_{S_+}(X_s)\,\mathrm{d}s / \mathbf{P}_x(\{\mathrm{D}_{S_+} < \infty\}), \qquad y \geq x.$$

Then for all $x \leq y$, $x, y \in J_n$, $n = 1, 2, \ldots$,

$$f_+(x,y) < \infty. \tag{5.9}$$

Clearly, for the proof of (5.9) we have only to consider the case $\mathbf{P}_x(\{\mathrm{D}_{S_+} < \infty\}) > 0$, $x \leq s_+(x) < y$, where $s_+(x)$ is defined in (2.12). But in this case (2.12)(i) yields $s_+(x) \in S_+$ and, therefore, $s_+(x) < y \in J_n$ implies $J_n \setminus R \subseteq K_+ \setminus E$. Combining (2.9) and (5.8) we obtain that

[2] Mind (1.28).

$$\begin{aligned} f_+(x,y) &= \mathbf{E}_x \int_{\mathrm{D}_{S_+}}^{\mathrm{D}_y} \mathbf{1}_{S_+}(X_s)\,\mathrm{d}s\,\mathbf{1}_{\{\mathrm{D}_{S_+}<\infty\}}/\mathbf{P}_x(\{\mathrm{D}_{S_+}<\infty\}) \\ &= \mathbf{E}_{s_+(x)} \int_0^{\mathrm{D}_y} \mathbf{1}_{S_+}(X_s)\,\mathrm{d}s \\ &\le \mathbf{E}_{s_+(x)}\mathrm{D}_y < \infty. \end{aligned}$$

Now we are able to prove the relation

$$\textbf{(5.10)} \qquad f_+(x,z) = f_+(x,y) + f_+(y,z), \qquad x \le y \le z,\ x,y,z \in J_n,$$

which is basic for the existence of the measure m_+^S.

Recalling the convention $0/0 = 0$ (cf. (1.28)) the strong Markov property of X yields

$$\begin{aligned} f_+(x,z) &= \frac{\mathbf{E}_x\left(\int_0^{\mathrm{D}_y} \mathbf{1}_{S_+}(X_s)\,\mathrm{d}s + \int_{\mathrm{D}_y}^{\mathrm{D}_z} \mathbf{1}_{S_+}(X_s)\,\mathrm{d}s\,\mathbf{1}_{\{\mathrm{D}_y<\infty\}}\right)}{\mathbf{P}_x(\{\mathrm{D}_{S_+}<\infty\})} \\ \textbf{(5.11)} \qquad &= f_+(x,y) + \frac{\mathbf{E}_y \int_0^{\mathrm{D}_z} \mathbf{1}_{S_+}(X_s)\,\mathrm{d}s\,\mathbf{P}_x(\{\mathrm{D}_y<\infty\})}{\mathbf{P}_x(\{\mathrm{D}_{S_+}<\infty\})}. \end{aligned}$$

In what follows we frequently use (2.12)(i).

If $x \le y \le s_+(x) \le z$ then we have

$$\begin{aligned} \mathbf{P}_x(\{\mathrm{D}_{S_+}<\infty\}) &= \mathbf{P}_x(\{\mathrm{D}_y<\infty, \mathrm{D}_{S_+}\circ\theta_{\mathrm{D}_y}<\infty\}) \\ &= \mathbf{P}_x(\{\mathrm{D}_y<\infty\})\mathbf{P}_y(\{\mathrm{D}_{S_+}<\infty\}) \end{aligned}$$

and (5.10) is a consequence of (5.11).

For $x \le s_+(x) < y \le z$ we have

$$\mathbf{P}_x(\{\mathrm{D}_y<\infty\}) = \mathbf{P}_x(\{\mathrm{D}_{s_+(x)}<\infty\})\mathbf{P}_{s_+(x)}(\{\mathrm{D}_y<\infty\})$$

which gives $\mathbf{P}_x(\{\mathrm{D}_y < \infty\}) = \mathbf{P}_x(\{\mathrm{D}_{s_+(x)} < \infty\})$ by using (5.8). Now in the case of $\mathbf{P}_x(\{\mathrm{D}_{s_+(x)} < \infty\}) = 0$ it follows from $x \le s_+(x) < y$ and $s_+(x) \in K_+ \setminus E$ that x and y are not connected which is impossible. Hence, $\mathbf{P}_x(\{\mathrm{D}_{s_+(x)} < \infty\}) > 0$ and thus $s_+(x) \in K_+ \setminus E$ which gives $\mathbf{P}_x(\{\mathrm{D}_y < \infty\}) = \mathbf{P}_x(\{\mathrm{D}_{S_+} < \infty\}) > 0$. Thus (5.11) leads to

$$f_+(x,z) = f_+(x,y) + \mathbf{E}_y \int_0^{\mathrm{D}_z} \mathbf{1}_{S_+}(X_s)\,\mathrm{d}s.$$

If the second summand vanishes, then (5.10) follows immediately. Otherwise, if $\mathbf{E}_y \int_0^{\mathrm{D}_z} \mathbf{1}_{S_+}(X_s)\,\mathrm{d}s > 0$ then we conclude $\mathbf{P}_y(\{\mathrm{D}_{S_+} < \infty\}) > 0$ and $s_+(y) < z$. Applying (5.8) again we compute

$$\begin{aligned} 1 &= \mathbf{P}_{s_+(x)}(\{\mathrm{D}_z<\infty\}) = \mathbf{P}_{s_+(x)}(\{\mathrm{D}_y<\infty, \mathrm{D}_z\circ\theta_{\mathrm{D}_y}<\infty\}) \\ &= \mathbf{P}_y(\{\mathrm{D}_z<\infty\}) \end{aligned}$$

establishing $\mathbf{P}_y(\{\mathrm{D}_{S_+} < \infty\}) = 1$, too. Therefore, the last equality for $f_+(x,z)$ coincides with (5.10).

Finally, for the remaining case $x \le y \le z \le s_+(x)$, (5.10) is obviously satisfied.

Now we are able to give the definition of m_+^S. For n fixed choose $x_0 \in J_n$ and set

$$\eta(y) = \begin{cases} f_+(x_0, y) & : \quad y \ge x_0,\ y \in J_n, \\ -f_+(y, x_0) & : \quad y \le x_0,\ y \in J_n. \end{cases}$$

Then η is increasing and continuous. Therefore, we can also define a measure $m_+^S(\cdot \cap J_n)$ on $(S_+, \mathfrak{B}(S_+))$ by

$$\textbf{(5.12)} \qquad m_+^S(A \cap J_n) = \int_{A \cap J_n} \eta(\mathrm{d}y).$$

Using (5.10) we see that $m_+^S(\cdot \cap J_n)$ does not depend on the choice of x_0.

Now we define the desired measure m_+^S as the sum

$$m_+^S(A) = \sum_n m_+^S(A \cap J_n), \qquad A \in \mathfrak{B}(S_+).$$

For all $y, z \in J_n$ with $y < z$ from this definition we have

$$m_+^S([y,z] \cap S_+) = \int_{[y,z]\cap S_+} \eta(\mathrm{d}x) = \int_{[y,z]\cap S_+} f_+(y, \mathrm{d}x).$$

To prove the assertion of (5.7) we have to verify that $m_+^S([y,z] \cap S_+) = f_+(y,z)$, $y, z \in J_n$, $y \le z$. In the case of $J_n \cap S_+ = \emptyset$ this relation is trivially true since both sides vanish. In the case of $J_n \cap S_+ \ne \emptyset$ we have $J_n \setminus S_+ \subseteq \bar{R}$ which implies

$$\begin{aligned} m_+^S([y,z] \cap S_+) &= \int_{[y,z]} f_+(y, \mathrm{d}x) - \int_{[y,z]\setminus S_+} f_+(y, \mathrm{d}x) \\ &= f_+(y,z) - \int_{[y,z]\cap \bar{R}} f_+(y, \mathrm{d}x). \end{aligned}$$

But it is easy to verify that the function $f_+(y,x)$, $x \ge y$, is constant on subintervals of $\bar{R}$. Altogether we get

$$m_+^S([y,z] \cap S_+) = f_+(y,z), \qquad y, z \in J_n,\ y \le z,$$

and this equality also yields the uniqueness of m_+^S. The σ-finiteness of m_+^S on each of the disjoint sets $J_n \cap S_+$ is due to (5.9) and so m_+^S is itself σ-finite.

Finally, the last assertion of the proposition follows from the continuity of $f_+(y, \cdot)$ on each interval J_n. □

Now all preparations are done to define the speed measure of X.

(5.13) Definition Let $(X, \mathbb{F})$ on $(\Omega, \mathcal{F}, \mathbf{P}_x, x \in I)$ be a continuous strong Markov process with scale function p. We call the measure m on $(I \setminus E, \mathfrak{B}(I \setminus E))$ satisfying

$$m|_{\bar{R} \setminus E} = m_r, \; m|_{S_+} = m_+^S, \; m|_{S_-} = m_-^S$$

the *speed measure of X associated with p.*

Observe that, given the process X and one of its scale functions, then the measure m is uniquely determined by (5.6)(i) and (5.7) justifying the above definition.

We provide some important properties of the speed measure in the next section after studying the occupation time formula.

V.2 Occupation Time Formula

In the present section we derive a useful formula concerning the so-called *occupation time* $\int_0^t \mathbf{1}_A(X_s)\,\mathrm{d}s$ which is the time the process X spends in a set $A \in \mathfrak{B}(I)$ up to the time t. This formula will play an important role in the next two chapters. It is used in Chapter VI to construct a continuous strong Markov process starting from a given scale function and a speed measure. In Chapter VII the occupation time formula will be applied to decide whether a continuous strong Markov process is a solution of a stochastic differential equation.

Let $(X, \mathbb{F})$ be a continuous strong Markov process with a fixed scale function p and associated speed measure m. First of all, we investigate the occupation time of X in certain subsets of $\bar{R}$, S_+ and S_-.

(5.14) Lemma *Let (c, d) denote an interval whose closure is included in a component (a, b) of R. We set $Y = p((X \vee c) \wedge d)$. Then it holds that*

$$\int_0^t \mathbf{1}_{(c,d)}(X_s)\,\mathrm{d}s = \int_{(c,d)} L^Y(t, p(y))\, m(\mathrm{d}y), \qquad t \geq 0, \; a.s.$$

Proof. Let q denote the inverse function of the restriction of p to (a, b). Further we set

$$\xi_t = \inf\{s \geq t : X_s \notin (c, d)\}, \qquad t \geq 0.$$

In view of (4.2)(i) and (1.17)(ii) we have $Y^{\xi_0} \in \mathcal{M}(\mathbb{F}^X)$. Choose a function f as in (5.1)(i). Since f is p-concave on (a, b) $f \circ q$ turns out to be concave on $(p(a+), p(b-))$. But we have $X^{\xi_0} = q(Y^{\xi_0})$ $\mathbf{P}_x$-a.s. for all $x \in [c, d]$ and, therefore, using the generalized Itô-formula (1.14), we can compute

$$\begin{aligned} f(X_t^{\xi_0}) &= f \circ q(Y_t^{\xi_0}) \\ &= f(X_0) + \int_0^t (f \circ q)'(Y_s^{\xi_0})\,\mathrm{d}Y_s^{\xi_0} + \\ &\quad + \frac{1}{2} \int_{[p(c), p(d)]} L^Y(t \wedge \xi_0, y)\, \mathrm{d}(f \circ q)'_+(y) \end{aligned}$$

for all $t \geq 0$ $\mathbf{P}_x$-a.s., $x \in [c, d]$. Here the stochastic integral with respect to Y^{ξ_0} is a continuous local $\mathbf{P}_x$-martingale for all $x \in [c, d]$ and the same holds true for $(f(X_t^{\xi_0}) + t \wedge \xi_0)_{t \geq 0}$ by (5.1)(i). Hence the process

$$\left(\frac{1}{2}\int_{[p(c),p(d)]} L^Y(t \wedge \xi_0, y)\, \mathrm{d}(f \circ q)'_+(y) \;+\; t \wedge \xi_0\right)_{t \geq 0}$$

is also a continuous local $\mathbf{P}_x$-martingale for all $x \in [c, d]$ and even for all $x \in I$ by the definition of ξ_0. The paths of the above process are of finite variation. It is well-known (cf. [12], Theorem T.V.39) that a continuous local martingale admitting paths of finite variation has to be constant. Thus[3]

$$-\frac{1}{2}\int_{(p(c),p(d))} L^Y(t \wedge \xi_0, y)\, \mathrm{d}(f \circ q)'_+(y) \;=\; t \wedge \xi_0, \qquad t \geq 0, \text{ a.s.},$$

which can also be written as

$$\int_{(c,d)} L^Y(t \wedge \xi_0, p(y))\, m(\mathrm{d}y) \;=\; t \wedge \xi_0, \qquad t \geq 0, \text{ a.s.},$$

using the definition (5.5) of the speed measure m on R.

We set

$$V_t \;=\; \int_{(c,d)} L^Y(t, p(y))\, m(\mathrm{d}y) \;-\; \int_0^t \mathbf{1}_{(c,d)}(X_s)\, \mathrm{d}s, \qquad t \geq 0,$$

and obtain from the above equality

$$\int_0^{\xi_0} |\mathrm{d}V_s| \;=\; 0 \qquad \text{a.s.}$$

By (1.16)(i), the process V is additive, and, consequently the Markov property of X leads to

$$\mathbf{P}_x(\{\int_t^{\xi_t} |\mathrm{d}V_s| = 0\}) \;=\; \mathbf{P}_x(\{\int_0^{\xi_0} |\mathrm{d}V_s| \circ \theta_t = 0\}) \;=\; 1, \qquad t \geq 0,\, x \in I.$$

In view of $\{t : t > 0,\, X_t \in (c, d)\} = \bigcup_{r \in \mathbb{Q}_+} (r, \xi_r)$ a.s. this implies

$$\int_0^\infty \mathbf{1}_{(c,d)}(X_s)\, |\mathrm{d}V_s| \;=\; 0 \qquad \text{a.s.}$$

or, in other words,

$$\int_0^t \mathbf{1}_{(c,d)}(X_s)\, \mathrm{d}s \;=\; \int_{(c,d)} L^Y(t, p(y))\, m(\mathrm{d}y), \qquad t \geq 0, \text{ a.s.}$$

proving the assertion. □

[3] We can omit the end-points of $[p(c), p(d)]$ by (1.17)(ii).

(5.15) Lemma *Let (a,b) be a component of R and assume that the lower end-point a belongs to $K_+ \setminus E$ and is attracting, i.e. $p(a) := p(a+) > -\infty$. Then we have*

$$\int_0^t \mathbf{1}_{\{a\}}(X_s)\,\mathrm{d}s = \frac{1}{2} L_+^Y(t, p(a)) m(\{a\}), \qquad t \geq 0, \text{ a.s.},$$

where $Y = p((X \vee a) \wedge d)$ for an arbitrary fixed point $d \in (a,b)$. An analogous result holds for an upper end-point $b \in K_- \setminus E$.

Proof. The proof is similar to the proof of (5.14) and so some details can be omitted.

We define

$$\xi_t = \inf\{s \geq t : X_s \notin [a,d)\}$$

and denote by q the inverse function of $p|_{[a,b)}$. If f_a is a function as in (5.1)(ii) then $f_a \circ q$ turns out to be concave on $[p(a), p(b))$ since f_a is p-concave on $[a,b)$. By (4.2)(i) we have that $Y^{\xi_0} \in \mathcal{S}(\mathbb{F}^X)$ and $X^{\xi_0} = q(Y^{\xi_0})$ holds $\mathbf{P}_x$-a.s. for all $x \in [a,d]$. Now from the generalized Itô-formula (1.14) (but now using the right local time) it follows that

$$f_a(X_t^{\xi_0}) = f_a(X_0) + \int_0^t (f_a \circ q)'_-(Y_s^{\xi_0})\,\mathrm{d}Y_s^{\xi_0} + \frac{1}{2} \int_{[p(a),p(d)]} L_+^Y(t \wedge \xi_0, y)\,\mathrm{d}(f_a \circ q)'_+(y)$$

for all $t \geq 0$ $\mathbf{P}_x$-a.s. for all $x \in [a,d]$. Thinking of $f_a \circ q$ as a function extended in its domain to $(-\infty, p(b))$ by setting $f_a \circ q(x) = f_a(a)$, $x < p(a)$, we have $(f_a \circ q)'_-(p(a)) = 0$. Therefore, using (4.2)(i) and (1.17)(ii), the stochastic integral in the above equation is a continuous local martingale. As in the proof of (5.14) we can deduce that

$$-\frac{1}{2} \int_{[p(a),p(d)]} L_+^Y(t \wedge \xi_0, y)\,\mathrm{d}(f_a \circ q)'_+(y) = t \wedge \xi_0, \qquad t \geq 0, \text{ a.s.}$$

Using (1.17)(iii) and the definition (5.13),(5.5) of $m(\{a\})$ for $a \in K_+ \setminus E$ this leads to

$$\begin{aligned} \int_0^{t\wedge\xi_0} \mathbf{1}_{\{a\}}(X_s)\,\mathrm{d}s &= \int_0^{t\wedge\xi_0} \mathbf{1}_{\{p(a)\}}(Y_s)\,\mathrm{d}s \\ &= -\frac{1}{2} L_+^Y(t \wedge \xi_0, p(a))\,(f_a \circ q)'_+(p(a)) \\ &= \frac{1}{2} L_+^Y(t \wedge \xi_0, p(a))\, m(\{a\}) \end{aligned}$$

for all $t \geq 0$ a.s. Finally, observe that because of $a_+ \in K_+ \setminus E$

$$\{t : t > 0,\ X_t \in [a,d)\} = \bigcup_{r \in \mathbb{Q}_+} (r, \xi_r) \qquad \text{a.s.}$$

and the assertion follows as in the proof of (5.14). □

(5.16) Lemma *For every set $A \in \mathfrak{B}(S_+)$ (resp. $A \in \mathfrak{B}(S_-)$) we have*

$$\int_0^t \mathbf{1}_A(X_s)\,ds = m([X_0, \max_{s\le t} X_s] \cap A), \qquad t \ge 0, \text{ a.s.},$$

$$(\textit{resp.} \int_0^t \mathbf{1}_A(X_s)\,ds = m([\min_{s\le t} X_s, X_0] \cap A), \qquad t \ge 0, \text{ a.s.}).$$

Proof. We only consider the case $A \in \mathfrak{B}(S_+)$. For $t < \mathrm{D}_{S_+}$ the assertion is trivial. For $\mathbf{P}_x(\{\mathrm{D}_{S_+} < \infty\}) > 0$ it remains to be proved that

$$\mathbf{P}_x(\{\int_0^{\mathrm{D}_{S_+}+t} \mathbf{1}_A(X_s)\,ds = m([X_0, \max_{s\le \mathrm{D}_{S_+}+t} X_s] \cap A),\ t \ge 0,\ \mathrm{D}_{S_+} < \infty\})$$

$$= \mathbf{P}_x(\{\mathrm{D}_{S_+} < \infty\}). \tag{5.17}$$

From (2.12)(i) it follows that

$$[X_0, \max_{s\le \mathrm{D}_{S_+}+t} X_s] = [x, s_+(x)) \cup [X_{\mathrm{D}_{S_+}}, \max_{\mathrm{D}_{S_+}\le s\le \mathrm{D}_{S_+}+t} X_s]$$

on $\{\mathrm{D}_{S_+} < \infty\}$ for all $t \ge 0$ $\mathbf{P}_x$-a.s. Further,

$$\int_0^{\mathrm{D}_{S_+}} \mathbf{1}_A(X_s)\,ds = 0 \quad \text{and} \quad m([x, s_+(x)) \cap A) = 0$$

holds. Therefore, applying the strong Markov property and (2.12)(i), (5.17) is equivalent to

$$\mathbf{P}_{s_+(x)}(\{\int_0^t \mathbf{1}_A(X_s)\,ds = m([X_0, \max_{s\le t} X_s] \cap A),\ t \ge 0\}) = 1.$$

Consequently, it suffices to deal with the case $x = s_+(x) \in S_+$, in what follows.

As in the proof of (5.7) we make use of the decomposition

$$I \setminus E = \bigcup_n J_n,\ J_n \cap J_m = \emptyset,\ n \ne m,$$

of $I \setminus E$ into intervals J_n with properties given in (2.8).

First, we verify the assertion in the case $A = S_+$. We can assume $x \in J_n \cap S_+$ implying $J_n \setminus R \subseteq K_+ \setminus E$ by (2.8). For $z \in J_n$, $z \ge x$, using (5.13),(5.7) and (4.18), we have

$$\int_0^{\mathrm{D}_z} \mathbf{1}_{S_+}(X_s)\,ds = \mathbf{E}_x \int_0^{\mathrm{D}_z} \mathbf{1}_{S_+}(X_s)\,ds = m([X_0, z] \cap S_+) \qquad \mathbf{P}_x - \text{a.s.}$$

Since the function $z \mapsto \mathrm{D}_z$ is $\mathbf{P}_x$-a.s. left continuous in $z \ge x$, $z \in J_n$, (see also (5.8)) both sides of this equality turn out to be $\mathbf{P}_x$-a.s. left continuous for $z \ge x$, $z \in J_n$. Consequently, it even holds that

$$\int_0^{\mathrm{D}_z} \mathbf{1}_{S_+}(X_s)\,\mathrm{d}s = m([X_0, z] \cap S_+), \qquad z \geq x,\ z \in J_n,\ \mathbf{P}_x - \text{a.s.},$$

that is, the $\mathbf{P}_x$-exceptional set does not depend on z. Denote

$$M_t := \max_{s \leq t} X_s = \inf\{z \geq x : \mathrm{D}_z > t\}, \qquad t \geq 0,$$

then from the last equality it follows that

$$\textbf{(5.18)} \quad \int_0^{t \wedge \mathrm{D}_z} \mathbf{1}_{S_+}(X_s)\,\mathrm{d}s = \int_0^{\mathrm{D}_{M_t \wedge z}} \mathbf{1}_{S_+}(X_s)\,\mathrm{d}s = m([X_0, M_t \wedge z] \cap S_+)$$

for all $t \geq 0$, $z \geq x$, $z \in J_n$, $\mathbf{P}_x$-a.s. Now let $y = \sup J_n$. By (2.8), if y belongs to I then it is a special singular point. This implies $X_t \notin S_+$, $t \geq \mathrm{D}_y$, $\mathbf{P}_x$-a.s., and thus

$$\int_0^t \mathbf{1}_{S_+}(X_s)\,\mathrm{d}s = \int_0^{t \wedge \mathrm{D}_y} \mathbf{1}_{S_+}(X_s)\,\mathrm{d}s,\ m([X_0, M_t] \cap S_+) = m([X_0, M_t \wedge y] \cap S_+)$$

for all $t \geq 0$ $\mathbf{P}_x$-a.s. Letting $z \uparrow y$ in (5.18) we finally obtain

$$\int_0^t \mathbf{1}_{S_+}(X_s)\,\mathrm{d}s = m([X_0, M_t] \cap S_+), \qquad t \geq 0,\ \mathbf{P}_x - \text{a.s.},$$

proving the assertion in the case of $A = S_+$. Finally, for an arbitrary $A \in \mathfrak{B}(S_+)$ we compute

$$\begin{aligned}
\int_0^t \mathbf{1}_A(X_s)\,\mathrm{d}s &= \int_0^t \mathbf{1}_A(X_s)\mathbf{1}_{S_+}(X_s)\,\mathrm{d}s \\
&= \int_0^t \mathbf{1}_A(X_s)\,\mathrm{d}m([X_0, M_s] \cap S_+) \\
&= \int_0^t \mathbf{1}_A(M_s)\,\mathrm{d}m([X_0, M_s] \cap S_+) \\
&= \int_{X_0}^{M_t} \mathbf{1}_{A \cap S_+}(y)\, m(\mathrm{d}y) \\
&= m([X_0, M_t] \cap A)
\end{aligned}$$

for all $t \geq 0$ a.s. finishing the proof. □

Now we combine (5.14), (5.15) and (5.16) into a general occupation time formula. For the first version of the occupation time formula we require $X \in \mathcal{S}(\mathbb{F}^X)$.

Let the function $\bar{g}$ be defined by (4.6).

(5.19) Theorem (Occupation time formula I) *Let X be a continuous strong Markov semimartingale. Then for every nonnegative measurable function $f : I \to [0,\infty)$ vanishing on E it holds that*

$$\begin{aligned}\int_0^t f(X_s)\,\mathrm{d}s &= \int_{\bar{R}\cap\{\bar{g}\neq 0\}\setminus E} L^X(t,y)(\bar{g}(y))^{-1} f(y)\,m(\mathrm{d}y) \\ &+ \int_0^t f(X_s)\mathbf{1}_{\bar{R}\cap\{\bar{g}=0\}}(X_s)\,\mathrm{d}s \\ &+ \int_{[X_0,\max_{s\le t} X_s]\cap S_+} f(y)\,m(\mathrm{d}y) \\ &+ \int_{[\min_{s\le t} X_s, X_0]\cap S_-} f(y)\,m(\mathrm{d}y)\end{aligned}$$

for all $t \geq 0$ a.s.

Proof. It suffices to prove the assertion for indicator functions $f = \mathbf{1}_A$, $A \in \mathfrak{B}(I)$, $A \subseteq (\bar{R} \cap \{\bar{g} \neq 0\} \setminus E) \cup S_+ \cup S_-$.

In the case of $A \subseteq S_+$ and $A \subseteq S_-$ we have already shown the formula in (5.16).

Let A be a measurable subset of an interval (c,d) whose closure is contained in a component (a,b) of R. Then (5.14) yields

$$\int_0^t \mathbf{1}_A(X_s)\,\mathrm{d}s = \int_A L^Y(t,p(y))\,m(\mathrm{d}y), \qquad t \geq 0, \text{ a.s.},$$

where $Y = p((X \vee c) \wedge d)$. As in the proof of (4.5) we can replace $L^Y(\cdot, p(y))$ by $L^X(\cdot,y)(\bar{g}(y))^{-1}$ for all $y \in (c,d) \cap \{\bar{g} \neq 0\}$ proving the assertion.

So the case $A = \{a\}$, $a \in K_+ \setminus E$ resp. $A = \{b\}$, $b \in K_- \setminus E$ remains to be considered. We restrict ourselves to the case $A = \{a\}$ by symmetry.

First let a be nonattracting for (a,b), i.e. $p(a+) = -\infty$. Then we have $\mathbf{P}_x(\{\mathrm{D}_a < \infty\}) = 0$ for all $a < x \in I$ (cf. (2.19)(iii)). Together with $a \in K_+ \setminus E$ this yields

$$\int_0^t \mathbf{1}_{\{a\}}(X_s)\,\mathrm{d}s = 0, \qquad t \geq 0, \mathbf{P}_x - \text{a.s.},$$

for all $x \in I$. On the other hand, $m(\{a\}) = -\frac{1}{2}\mathrm{D}_p^+ f_a(a)$ vanishes because of $p(a+) = -\infty$ proving the formula in that case.

Now let a be attracting for (a,b). We set $p(a) = p(a+) > -\infty$ and conclude from (5.15) that

$$\int_0^t \mathbf{1}_{\{a\}}(X_s)\,\mathrm{d}s = \frac{1}{2} L_+^Y(t,p(a))\,m(\{a\}), \qquad t \geq 0, \text{ a.s.},$$

where $Y = p((X \vee a) \wedge d)$ with $d \in (a,b)$. Finally, in the proof of (4.7) we have verified the equality

(5.20) $$L^X(\cdot, a) = \frac{1}{2} L_+^Y(\cdot, p(a))\bar{g}(a) \text{ a.s.}$$

leading to

$$\int_0^t \mathbf{1}_{\{a\}}(X_s)\, ds = L^X(t,a)(\bar{g}(a))^{-1} m(\{a\}), \qquad t \geq 0, \text{ a.s.},$$

if $\bar{g}(a) \neq 0$. □

In the above formula we were not able to describe the occupation time of X in measurable subsets of $\bar{R} \cap \{\bar{g} = 0\}$ by the local time L^X. This is due to the fact that $L^X(t, \cdot)$ just vanishes on $\bar{R} \cap \{\bar{g} = 0\}$ (cf. (4.7)) while $\int_0^t f(X_s)\mathbf{1}_{\bar{R}\cap\{\bar{g}=0\}}(X_s)\, ds$ does not need to vanish. However, assuming an additional property of the scale function p of X which is not a vital restriction it is possible to express the occupation time of X in measurable subsets of $\bar{R}$ completely by the local time of the process $p(X)$.

For the remaining part of this chapter let us assume that each end-point of a component (a, b) of R which is an element of I is attracting for (a, b), i.e.,

$$\text{if } a \in I \text{ then } p(a+) = \lim_{x \downarrow a} p(x) > -\infty$$

and

$$\text{if } b \in I \text{ then } p(b-) = \lim_{x \uparrow b} p(x) < \infty$$

for every scale function p of X. Therefore, because a scale function is only unique up to an affine transformation on the components of R, we may choose a scale function p of X satisfying

$$\sum_{n: a_n \in I, b_n \in I} (p(b_n-) - p(a_n+)) < \infty.$$

Fix $x_0 \in \text{int}(I)$. Then,

$$\tilde{p}(x) = \begin{cases} \sum_n \int_{(a_n,b_n)\cap[x_0,x)} p(dy) + \int_{I\cap[x_0,x)\setminus R} dy & : \; x \geq x_0 \\ \sum_n \int_{(a_n,b_n)\cap(x,x_0]} p(dy) + \int_{I\cap(x,x_0]\setminus R} dy & : \; x < x_0 \end{cases}$$

is a strictly increasing continuous function from I into $\mathbb{R}$ whose restriction to R is a scale function of X. Thus, the above assumption on p turns out to be equivalent to the following condition:

(5.21) There exists a strictly increasing continuous function $p : I \to \mathbb{R}$ whose restriction $p|_R$ to R is a scale function of X.

Whenever (5.21) is assumed, the function p there is also called a scale function of X.

(5.22) Lemma *Let* $(X, \mathbb{F})$ *on* $(\Omega, \mathcal{F}, \mathbf{P}_x, x \in I)$ *be a continuous strong Markov process with a scale function* p *according to (5.21). Then,* $p(X) \in \mathcal{S}(\mathbb{F}^X)$ *holds and* $(p(X), \mathbb{F})$ *is strong Markov with respect to* $(\Omega, \mathcal{F}, \mathbf{P}_{q(x)}, x \in p(I))$ *where* q *denotes the inverse function to* p.

Proof. The strong Markov property of $(p(X), \mathbb{F})$ with respect to $(\Omega, \mathcal{F}, \mathbf{P}_{q(x)}, x \in p(I))$ is evident. It remains to be shown that $p(X) \in \mathcal{S}(\mathbb{F}^X)$.

Setting $Y = p(X)$ we have[4]

$$\begin{aligned} Y &= Y_0 + \ell([Y_0, Y] \cap p(S_+)) - \ell([Y, Y_0] \cap p(S_+)) \\ &\quad + \ell([Y_0, Y] \cap p(S_-)) - \ell([Y, Y_0] \cap p(S_-)) \\ &\quad + \ell([Y_0, Y] \cap p(\bar{R})) - \ell([Y, Y_0] \cap p(\bar{R})). \end{aligned}$$

But (2.12) gives $\ell([Y, Y_0] \cap p(S_+)) = 0$ as well as $\ell([Y_0, Y] \cap p(S_-)) = 0$ a.s. implying

$$\begin{aligned} Y &= Y_0 + \ell([Y_0, Y] \cap p(S_+)) - \ell([Y, Y_0] \cap p(S_-)) + \\ &\quad \sum_n (\ell([Y_0, Y] \cap p((a_n, b_n))) - \ell([Y, Y_0] \cap p((a_n, b_n)))) \\ &= Y_0 + \ell([Y_0, Y] \cap p(S_+)) - \ell([Y, Y_0] \cap p(S_-)) + \\ &\quad \sum_n ((Y \vee p(a_n)) \wedge p(b_n) - (Y_0 \vee p(a_n)) \wedge p(b_n)) \text{ a.s.} \end{aligned}$$

We only have to prove that the last sum belongs to $\mathcal{S}(\mathbb{F}^X)$ since, by $S_+ \subseteq K_+$, $S_- \subseteq K_-$ and (2.12), the processes $\ell([Y_0, Y] \cap p(S_+))$ and $\ell([Y, Y_0] \cap p(S_-))$ are increasing. We define

$$Y^n = (Y \vee p(a_n)) \wedge p(b_n)$$

and conclude from (4.2)(i) that $Y^n \in \mathcal{S}(\mathbb{F}^X)$ and

(5.23) $$Y^n - Y_0^n = M^n + L^{Y^n}(\cdot, p(a_n)) - L^{Y^n}(\cdot, p(b_n))$$

where $M^n \in \mathcal{M}(\mathbb{F}^X)$. Hence, in the case of only a finite number n_0 of components (a_n, b_n) of $R = \bigcup_{n=1}^{n_0} (a_n, b_n)$ we immediately obtain

$$\sum_{n=1}^{n_0} ((Y \vee p(a_n)) \wedge p(b_n) - (Y_0 \vee p(a_n)) \wedge p(b_n)) = \sum_{n=1}^{n_0} (Y^n - Y_0^n) \in \mathcal{S}(\mathbb{F}^X).$$

Now suppose that R possesses infinitely many components. We will verify the semimartingale property of

$$(\sum_n (Y^n - Y_0^n), \mathbb{F}^X)$$

[4] Here ℓ denotes the Lebesgue measure on $\mathbb{R}$.

with respect to $\mathbf{P}_x$ for every $x \in I$ proving the assertion. Using the localizing sequence of stopping times

$$\tau_k = \inf\{s \geq 0 : |Y_s - Y_0| \geq k\}, \qquad k = 1, 2, ...,$$

we obtain $\mathbf{P}_x$-a.s.

$$(5.24) \qquad |Y^n_{\cdot \wedge \tau_k} - Y^n_0| \leq \ell((p(a_n), p(b_n)) \cap (p(x) - k, p(x) + k))$$

and, hence,

$$\sum_n |Y^n_{\cdot \wedge \tau_k} - Y^n_0| \leq 2k \qquad \mathbf{P}_x - \text{a.s.}$$

holds. That is, for each $k = 1, 2, ...$, the above series is absolutely convergent $\mathbf{P}_x$-a.s. Let $\mathbb{K}$ denote the set $\{n : a_n \in K_+, b_n \in K_-\}$. Then

$$\sum_n (Y^n_{\cdot \wedge \tau_k} - Y^n_0) = \sum_{n \in \mathbb{K}} (Y^n_{\cdot \wedge \tau_k} - Y^n_0) + \sum_{n \in \mathbb{N} \setminus \mathbb{K}} (Y^n_{\cdot \wedge \tau_k} - Y^n_0),$$

and, with respect to $\mathbf{P}_x$, then the first sum on the right-hand side consists of at most two nonvanishing summands. Therefore $(\sum_{n \in \mathbb{K}} (Y^n_{\cdot \wedge \tau_k} - Y^n_0), \mathbb{F}^X)$ is a $\mathbf{P}_x$-semimartingale.

We investigate the remaining sum

$$(5.25) \qquad \sum_{n \in \mathbb{N} \setminus \mathbb{K}} (Y^n_{\cdot \wedge \tau_k} - Y^n_0)$$

and show its convergence according to a norm which is known to preserve the semimartingale property.

Let $(Z, \mathbb{F})$ be a continuous semimartingale on a probability space $(\Omega, \mathcal{F}, \mathbf{P})$ with canonical decomposition $Z = Z_0 + M + V$. Then Z belongs to the space $\mathcal{H}^1$ (see [14], VII.98) if

$$||Z|| := \mathbf{E}\left(|Z_0| + \langle M \rangle^{\frac{1}{2}}_\infty + \int_0^\infty |\mathrm{d}V| \right) < \infty.$$

We set $Z^*_t = \sup_{s \leq t} |Z_s|$ and apply the Burkholder-Davis-Gundy-inequality to obtain the estimate

$$\begin{aligned} ||Z|| &\leq c\mathbf{E}(|Z_0| + M^*_\infty + \int_0^\infty |\mathrm{d}V|) \\ (5.26) \qquad &\leq 2c\mathbf{E}((Z - Z_0)^*_\infty + \int_0^\infty |\mathrm{d}V|) \end{aligned}$$

with some constant $c > 0$.

Now we are going to prove the convergence of the series (5.25) with respect to the norm $|| \cdot ||$. To this end we estimate the $|| \cdot ||$-norm of $Y^n_{\cdot \wedge \tau_k} - Y^n_0$ for $n \in \mathbb{N} \setminus \mathbb{K}$. First, (5.24) implies

$$\mathbf{E}_x (Y^n_{\cdot \wedge \tau_k} - Y^n_0)^*_\infty \leq \ell((p(a_n), p(b_n)) \cap (p(x) - k, p(x) + k)).$$

Further, since $n \in \mathbb{N}\setminus\mathbb{K}$, at least one of the two local times in (5.23) vanishes. We restrict ourselves to the case $a_n \in K_+$, $b_n \in K_+$. Then

$$Y^n_{\cdot\wedge\tau_k} - Y^n_0 = M^n_{\cdot\wedge\tau_k} + L^{Y^n}(\cdot\wedge\tau_k, p(a_n)).$$

Setting $V^n := L^{Y^n}(\cdot\wedge\tau_k, p(a_n))$ from (5.24) we obtain

$$\begin{aligned}
\mathbf{E}_x \int_0^\infty |\mathrm{d}V^n| &= \lim_{t\uparrow\infty} \mathbf{E}_x L^{Y^n}(t\wedge\tau_k, p(a_n)) \\
&= \lim_{t\uparrow\infty} \mathbf{E}_x (Y^n_{t\wedge\tau_k} - Y^n_0 - M^n_{t\wedge\tau_k}) \\
&\leq \ell((p(a_n), p(b_n)) \cap (p(x)-k, p(x)+k)).
\end{aligned}$$

Using (5.26) this leads to

$$\|Y^n_{\cdot\wedge\tau_k} - Y^n_0\| \leq 4c\ell((p(a_n), p(b_n)) \cap (p(x)-k, p(x)+k))$$

and, consequently,

$$\sum_{n\in\mathbb{N}\setminus\mathbb{K}} \|Y^n_{\cdot\wedge\tau_k} - Y^n_0\| \leq 8ck.$$

Therefore, the series (5.25) converges in the norm $\|\cdot\|$ of the space $\mathcal{H}^1$ which is known to be a complete space of semimartingales (see [14], VII.98). Finally, since τ_k tends to infinity $\mathbf{P}_x$-a.s.,

$$\left(\sum_n (Y^n - Y^n_0), \mathbb{F}^X\right)$$

is a $\mathbf{P}_x$-semimartingale. □

Now, under the assumption (5.21) we are able to derive an "improved" occupation time formula. Contrary to the occupation time formula (5.19), now we do not require that $X \in \mathcal{S}(\mathbb{F}^X)$.

(5.27) Theorem (Occupation time formula II) *Let $(X, \mathbb{F})$ be a continuous strong Markov process with a scale function according to (5.21) and a speed measure m associated with p. Then, for every nonnegative measurable function $f: I \to [0,\infty)$ vanishing on E it holds that*

$$\begin{aligned}
\int_0^t f(X_s)\,\mathrm{d}s &= \int_{\bar{R}\setminus E} L^{p(X)}(t, p(y)) f(y)\, m(\mathrm{d}y) \\
&\quad + \int_{[X_0, \max_{s\leq t} X_s]\cap S_+} f(y)\, m(\mathrm{d}y) \\
&\quad + \int_{[\min_{s\leq t} X_s, X_0]\cap S_-} f(y)\, m(\mathrm{d}y)
\end{aligned}$$

for all $t \geq 0$ a.s.

Proof. Let (a_n, b_n) denote a component of R and let

$$Y^n = p((X \vee a_n) \wedge b_n) = (p(X) \vee p(a_n)) \wedge p(b_n).$$

Then, applying (1.17)(v), we obtain

$$L_{\pm}^{Y^n}(\cdot, p(y)) = L_{\pm}^{p(X)}(\cdot, p(y)) \qquad \text{a.s.}$$

for every $y \in (a_n, b_n)$. Consequently, using the properties of continuity of the local time (see (1.12)), it even holds that

$$\textbf{(5.28)} \qquad L_{\pm}^{Y^n}(\cdot, p(y)) = L_{\pm}^{p(X)}(\cdot, p(y))$$

for all $y \in (a_n, b_n)$ a.s., that is the exceptional set does not depend on y.

(5.14) implies

$$\int_0^t \mathbf{1}_{(a_n,b_n)}(X_s)\,\mathrm{d}s = \int_{(a_n,b_n)} L^{Y^n}(t, p(y))\, m(\mathrm{d}y), \qquad t \geq 0, \text{ a.s.,}$$

and, by (5.28) and (1.17)(iii),

$$\int_0^t \mathbf{1}_A(X_s)\,\mathrm{d}s = \int_A L^{p(X)}(t, p(y))\, m(\mathrm{d}y), \qquad t \geq 0, \text{ a.s.,}$$

for every measurable subset A of (a_n, b_n). Further, for $a_n \in K_+ \setminus E$ (5.15) implies

$$\int_0^t \mathbf{1}_{\{a_n\}}(X_s)\,\mathrm{d}s = \frac{1}{2} L_+^{Y^n}(t, p(a_n)) m(\{a_n\}), \qquad t \geq 0, \text{ a.s.}$$

But, applying Proposition (1.17)(v) again we obtain

$$L_+^{Y^n}(\cdot, p(a_n)) = L_+^{p(X)}(\cdot, p(a_n)) \qquad \text{a.s.}$$

while (1.17)(iv) yields

$$L_-^{p(X)}(\cdot, p(a_n)) = 0 \qquad \text{a.s.}$$

since $a_n \in K_+$. Therefore, from the definition of the symmetric local time it follows that

$$\textbf{(5.29)} \qquad \frac{1}{2} L_+^{Y^n}(\cdot, p(a_n)) = L^{p(X)}(\cdot, p(a_n)), \text{ a.s.,}$$

leading to

$$\int_0^t \mathbf{1}_{\{a_n\}}(X_s)\,\mathrm{d}s = L^{p(X)}(t, p(a_n)) m(\{a_n\}), \qquad t \geq 0, \text{ a.s.}$$

An analogous result is true for $b_n \in K_- \setminus E$.

Combining the above part of the proof and (5.16) the asserted equality follows. □

The next proposition shows that the occupation time formula in (5.27) already determines the speed measure of X associated with p.

(5.30) Proposition *Let $(X, \mathbb{F})$ be a continuous strong Markov process with a scale function p according to (5.21). If n is a measure on $(I \setminus E, \mathfrak{B}(I \setminus E))$ satisfying*

$$(5.31) \qquad t = \int_{\tilde{R}\setminus E} L^{p(X)}(t, p(y))\, n(dy) + \\ + \; n([X_0, \max_{s\le t} X_s] \cap S_+) + n([\min_{s\le t} X_s, X_0] \cap S_-)$$

for all $t < \mathrm{D}_E$ a.s. then n is the speed measure of X associated with p.

Remark. Observe that (5.31) is actually equivalent to the conclusion of (5.27) with m replaced by n. Indeed, this follows from (1.17)(iii), the definitions of S_+, S_- and (2.14).

Proof. Let m denote the speed measure of X associated with p. First, we verify $n|_{\tilde{R}\setminus E} = m|_{\tilde{R}\setminus E}$. From our assumption and (1.17)(iii),

$$(5.32) \qquad \int_B L^{p(X)}(t, p(y))\, n(dy) = \int_B L^{p(X)}(t, p(y))\, m(dy)$$

is true for all $t \ge 0$ $\mathbf{P}_x$-a.s. and every $B \in \mathfrak{B}(\tilde{R} \setminus E)$. In the proof of (4.7) we have shown that

$$\mathbf{P}_y(\{L^{p(X)}(t, p(y)) > 0\}) = 1, \qquad t > 0,\ y \in R.$$

Moreover, combining (5.28), (4.2)(i) and Thèoréme 2 from [51], we obtain that $L^{p(X)}(t, p(\cdot))$ is continuous on R for every $t \ge 0$.

Now we fix a compact subinterval K of R. Then, due to (5.6)(iii), $\eta = n|_K - m|_K$ is a signed measure on $(K, \mathfrak{B}(K))$.

Let $\eta = \eta_1 - \eta_2$, $\eta_1 = \eta(\cdot \cap A_1)$, $\eta_2 = \eta(\cdot \cap A_2)$, $A_1 \cup A_2 = K$, $A_1 \cap A_2 = \emptyset$ be the Jordan-decomposition of η, that is η_1 and η_2 are nonnegative measures on $(K, \mathfrak{B}(K))$. Then, from (5.32) we have that

$$0 = \int_{A_i} L^{p(X)}(t, p(y))\, \eta(dy) = \int_K L^{p(X)}(t, p(y))\, \eta_i(dy), \qquad t \ge 0, \text{ a.s.},$$

for $i = 1, 2$ and, using the properties of $L^{p(X)}(t, p(\cdot))$ stated above, it follows that $\eta_i = 0$, $i = 1, 2$, i.e. $n|_K = m|_K$. But this means

$$n|_R = m|_R.$$

Now, let (a, b) denote a component of R with $a \in K_+ \setminus E$ (resp. $b \in K_- \setminus E$). Let $K = [a, a + \varepsilon]$ or $K = [b - \varepsilon, b]$, respectively, with $\varepsilon > 0$ sufficiently small. As above one verifies that $n|_K = m|_K$ but now using (5.6)(iv), (5.29), the right (resp. left) continuity of the local time on K and $\mathbf{P}_a(\{L^{p(X)}(t, p(a)) > 0\}) = 1$ (resp. $\mathbf{P}_b(\{L^{p(X)}(t, p(b)) > 0\}) = 1$), $t > 0$, (cf. the proof of (4.7)).

Summarizing,

$$n|_{\bar{R}\setminus E} = m|_{\bar{R}\setminus E}.$$

holds.

Now it remains to be verified that $n|_{S_+} = m|_{S_+}$ and $n|_{S_-} = m|_{S_-}$. To this end we show that n satisfies the equalities stated in (5.7). Fix $y, z \in J_n$, $y < z$, and assume $[y, z] \cap S_+ \neq \emptyset$. The points y and z are connected (cf. (2.8)) and, therefore, we have $\mathbf{P}_y(\{\mathrm{D}_{S_+} < \infty\}) > 0$. Combining (5.31), (5.8) and (2.12) we now obtain

$$\begin{aligned} n([y,z] \cap S_+) &= n([s_+(y), z] \cap S_+) \\ &= \mathbf{E}_{s_+(y)} \int_0^{\mathrm{D}_z} \mathbf{1}_{S_+}(X_s)\,\mathrm{d}s \\ &= \mathbf{E}_y \int_0^{\mathrm{D}_z} \mathbf{1}_{S_+}(X_s)\,\mathrm{d}s / \mathbf{P}_y(\{\mathrm{D}_{S_+} < \infty\}) \end{aligned}$$

proving $n|_{S_+} = m_+^S = m|_{S_+}$. Analogously one can verify that $n|_{S_-} = m|_{S_-}$. □

(5.33) Corollary *Let $(X, \mathbb{F})$ be a continuous strong Markov process with a scale function p according to (5.21) and a speed measure m associated with p. Further, let $q : p(I) \to I$ denote the inverse function to p. Then the function $i(x) = x$, $x \in p(I)$, is a scale function for the continuous strong Markov semimartingale $p(X)$ and $m \circ q$ is the associated speed measure.*

Proof. First, the property (2.17) defining a scale function is easily verified for the function i with respect to $p(X)$ and therefore omitted.

Let us deal with the speed measure of $p(X)$ associated with i. Let f be a function $f : p(I) \to [0, \infty)$ vanishing on $p(E)$. Then, by (5.27) and the transformation theorem for measures,

$$\begin{aligned} \int_0^t f(p(X_s))\,\mathrm{d}s &= \int_{p(\bar{R}\setminus E)} L^{p(X)}(t, y) f(y)\; m \circ q(\mathrm{d}y) \\ &+ \int_{[p(X_0), \max_{s\le t} p(X_s)] \cap p(S_+)} f(y)\; m \circ q(\mathrm{d}y) \\ &+ \int_{[\min_{s \le t} p(X_s), p(X_0)] \cap p(S_-)} f(y)\; m \circ q(\mathrm{d}y) \end{aligned}$$

for all $t \geq 0$ $\mathbf{P}_x$-a.s. For the process $p(X)$ the set of regular, right-singular, left-singular or absorbing points is $p(R)$, $p(K_+)$, $p(K_-)$ and $p(E)$, respectively. Applying (5.30) for $p(X)$ and i, the assertion follows. □

At the end of this chapter we collect some important properties of the speed measure.

(5.34) Proposition *The speed measure m of X associated with the scale function p possesses the following properties:*

(i) $m(G) > 0$ *for every open subset* $G \subseteq I \setminus E$.

(ii) $m(K) < \infty$ *for every compact subinterval* $K \subseteq R$.

(iii) $m(\{x\}) = 0$ *for every* $x \in S_+ \cup S_-$.

(iv) m *is σ-finite.*

If we suppose (5.21), additionally, then it holds that

(v) *For all* $x \in K_+ \setminus E$ *resp.* $x \in K_- \setminus E$ *there exists an* $\varepsilon > 0$, *such that*

$$\sum_n \int_{[a_n,b_n)\cap[x,x+\varepsilon)} (p(b_n \wedge (x+\varepsilon)) - p(y))\, m(dy) + m([x,x+\varepsilon)\cap S_+) < \infty$$

resp.

$$\sum_n \int_{(a_n,b_n]\cap(x-\varepsilon,x]} (p(y) - p(a_n \vee (x-\varepsilon)))\, m(dy) + m((x-\varepsilon,x]\cap S_-) < \infty.$$

(vi) *For all* $x \in I$ *satisfying* $[x,\infty) \cap K_- = \emptyset$ *resp.* $(-\infty,x] \cap K_+ = \emptyset$ *it holds that*

$$\int_{I\cap[x,\infty)} \sum_n \mathbf{1}_{[a_n,b_n)}(y)(p(b_n) - p(y))\, m(dy) + m([x,\infty)\cap S_+) = \infty$$

resp.

$$\int_{I\cap(-\infty,x]} \sum_n \mathbf{1}_{(a_n,b_n]}(y)(p(y) - p(a_n))\, m(dy) + m((-\infty,x]\cap S_-) = \infty.$$

Proof. (i) Let G be an open subset of $I \setminus E$. In the case of $G \cap \bar{R} \neq \emptyset$ we also have $G \cap R \neq \emptyset$ and the assertion follows from (5.6)(ii). Now, we suppose that $G \cap (\bar{R} \setminus E) = \emptyset$ as well as $G \cap S_+ \neq \emptyset$ without loss of generality. For $x \in G \cap S_+$ then there exists an $\varepsilon > 0$ such that $[x, x+\varepsilon) \subseteq G \cap S_+$. Indeed, otherwise we would have $[x, x+\varepsilon) \cap K_- \neq \emptyset$ for all $\varepsilon > 0$ implying the contradiction $x \in K_- \cap S_+$ by (2.5)(i). As a consequence, (5.19) yields

$$0 < \mathbf{E}_x \mathrm{D}_{x+\varepsilon} = \mathbf{E}_x \int_0^{\mathrm{D}_{x+\varepsilon}} \mathbf{1}_{[x,x+\varepsilon)}(X_s)\, ds = m([x, x+\varepsilon)) \leq m(G).$$

(ii) is nothing else but (5.6)(iii).

The property (iii) follows from (5.7) and (5.13).

(iv) Combining (ii) and (5.6)(iv) we obtain that $m|_{\bar{R}\setminus E}$ is σ-finite while for $m|_{S_+}$ and $m|_{S_-}$ this property has been shown in (5.7).

(v) By symmetry it suffices to treat the case $x \in K_+ \setminus E$. Here there exists a $z > x$ with $\mathbf{P}_x(\{\mathrm{D}_z < \infty\}) > 0$ and (2.9) even gives $\mathbf{E}_x \mathrm{D}_z < \infty$. Obviously, we also have $[x, z) \cap K_- = \emptyset$ leading to

$$\infty > \mathbf{E}_x \mathrm{D}_z = \mathbf{E}_x \int_0^{\mathrm{D}_z} \mathbf{1}_{[x,z)}(X_s)\,\mathrm{d}s$$

$$(5.35) \qquad = \mathbf{E}_x \int_{[x,z)\cap \bar{R}} L^{p(X)}(\mathrm{D}_z, p(y))\, m(\mathrm{d}y) + m([x,z)\cap S_+)$$

by (5.27). For $y \in [x,z) \cap \bar{R} \neq \emptyset$ there exists a component (a_n, b_n) of R with $a_n \in K_+ \setminus E$ such that $y \in [a_n, b_n \wedge z)$. We show that

$$(5.36) \quad p(b_n \wedge z) - p(y) \leq \mathbf{E}_x L^{p(X)}(\mathrm{D}_z, p(y)) \leq 2(p(b_n \wedge z) - p(y))$$

for all $y \in [a_n, b_n \wedge z)$. Combining (5.35) and the left-hand inequality of (5.36), it is then easy to deduce the assertion of item (v).

In order to prove (5.36) we observe that

$$\mathbf{E}_x L^{p(X)}(\mathrm{D}_z, p(y)) = \mathbf{E}_{a_n} L^{p(X)}(\mathrm{D}_z, p(y)), \qquad y \in [a_n, b_n \wedge z),$$

by the strong Markov property and (1.16)(i), (1.17)(i). If $b_n < z$ then $b_n \in K_+ \setminus E$ and, therefore, it even holds that

$$\mathbf{E}_{a_n} L^{p(X)}(\mathrm{D}_z, p(y)) = \mathbf{E}_{a_n} L^{p(X)}(\mathrm{D}_{z\wedge b_n}, p(y)), \qquad y \in [a_n, b_n \wedge z).$$

Now, we use the generalized Itô-formula (1.14) to compute

$$\mathbf{E}_{a_n} \frac{1}{2} L_+^{p(X)}(\mathrm{D}_{z\wedge b_n}, p(y)) = \mathbf{E}_{a_n} \left(p(X_{\mathrm{D}_{z\wedge b_n}}) - p(y)\right)^+ - (p(a_n) - p(y))^+ - \mathbf{E}_{a_n} \int_0^{\mathrm{D}_{z\wedge b_n}} \mathbf{1}_{(p(y),\infty)}(p(X_s))\,\mathrm{d}p(X_s)$$

$$(5.37) \qquad = p(b_n \wedge z) - p(y) - \mathbf{E}_{a_n} \int_0^{\mathrm{D}_{z\wedge b_n}} \mathbf{1}_{(p(y),\infty)}(p(X_s))\,\mathrm{d}p(X_s).$$

But from (5.23) and (1.17)(ii) we know that the $\mathbf{P}_{a_n}$-semimartingale decomposition of $p(X)$ up to $\mathrm{D}_{z\wedge b_n}$ is

$$p(X_t) = p(a_n) + M_t^n + L^{p(X)}(t, p(a_n)), \qquad t \leq \mathrm{D}_{z\wedge b_n}, \ \mathbf{P}_{a_n} - \text{a.s.}$$

Now (5.37) and (1.17)(iii) imply

$$\mathbf{E}_{a_n} \frac{1}{2} L_+^{p(X)}(\mathrm{D}_{z\wedge b_n}, p(y)) = p(b_n \wedge z) - p(y), \qquad y \in [a_n, b_n \wedge z).$$

Using Thèoréme 2 from [51], we see that $L^{p(X)}(t, p(\cdot))$ is continuous on R, $t \geq 0$, a.s., implying $L_+^{p(X)}(\cdot, p(y)) = L^{p(X)}(\cdot, p(y))$ a.s. for $y \in (a_n, b_n \wedge z)$. For $y = a_n \in K_+ \setminus E$, $\frac{1}{2} L_+^{p(X)}(\cdot, p(a_n)) = L^{p(X)}(\cdot, p(a_n))$ holds a.s. (see also (5.29)). Altogether, this gives (5.36).

(vi) Again we only consider the first case $[x,\infty) \cap K_- = \emptyset$. First, let $x \in K_+ \setminus E$. Then, for a sequence $x < z_k < \sup I$, $z_k \uparrow \sup I$, the assumption

$[x,\infty)\cap K_- = \emptyset$ implies $\mathbf{P}_x(\{\lim_{k\to\infty} \mathrm{D}_{z_k} = \infty\}) = 1$. Combining (5.27) and the right-hand inequality of (5.36) we obtain that

$$\begin{aligned}
\infty &= \lim_{k\to\infty} \mathbf{E}_x \mathrm{D}_{z_k} \\
&= \lim_{k\to\infty} \mathbf{E}_x \int_0^{\mathrm{D}_{z_k}} \mathbf{1}_{[x,\infty)\cap I}(X_s)\,\mathrm{d}s \\
&\le \int_{I\cap[x,\infty)} 2\sum_n \mathbf{1}_{[a_n,b_n)}(y)(p(b_n)-p(y))\,m(\mathrm{d}y) \;+\; m([x,\infty)\cap S_+).
\end{aligned}$$

Now, let $x \in (a_n, b_n) \subseteq R$. In the case of $b_n < \sup I$ from the assumptions it follows that $b_n \in K_+ \setminus E$ and we can argue for b_n as above for $x \in K_+ \setminus E$. So the case $b_n = \sup I$ remains to be considered. Here we assume $p(b_n) < \infty$ in what follows because (vi) is obviously satisfied for $p(b_n) = \infty$. Then, b_n cannot be an element of I because otherwise $b_n \in [x,\infty) \cap K_-$ which is a contradiction of our assumption. Therefore, again we have $\mathbf{P}_x(\{\lim_{k\to\infty} \mathrm{D}_{z_k} = \infty\}) = 1$ for sequences $I \ni z_k \uparrow b_n$. Applying (2.17) for $a_n < u < x < z_k < b_n$ we compute

$$\begin{aligned}
\mathbf{P}_x(\{\lim_{k\to\infty} \mathrm{D}_{z_k} \le \mathrm{D}_u\}) &= \lim_{k\to\infty} \mathbf{P}_x(\{\mathrm{D}_{z_k} \le \mathrm{D}_u\}) \\
&= (p(x)-p(u))/(p(b_n)-p(u)) \\
&> 0.
\end{aligned}$$

Consequently, for $A = \{\lim_{k\to\infty} \mathrm{D}_{z_k} \le \mathrm{D}_u\}$ it holds that

$$\lim_{k\to\infty} \mathbf{E}_x \mathrm{D}_{z_k} \mathbf{1}_A = \infty.$$

Finally, from (5.27) and (5.36) we conclude that

$$\begin{aligned}
\infty &= \lim_{k\to\infty} \mathbf{E}_x \mathrm{D}_{z_k} \mathbf{1}_A \\
&\le \lim_{k\to\infty} \mathbf{E}_x \int_0^{\mathrm{D}_{z_k}} \mathbf{1}_{[u,b_n)}(X_s)\,\mathrm{d}s \\
&\le 2\int_{[u,b_n)} (p(b_n)-p(y))\,m(\mathrm{d}y).
\end{aligned}$$

In view of $m([u,x]) < \infty$ (see (ii)) this proves the assertion. □

VI. Construction of Continuous Strong Markov Processes

In this chapter we establish a general method for the construction of continuous strong Markov processes. The only but not so vital restriction we make for technical reasons is (5.21).

The basic idea consists of the following. Starting from a given Wiener process W and a suitable weakly additive functional A of W we change the time of W by the right-inverse process of A followed by a space transformation to obtain the desired continuous strong Markov process. Here we especially use the results of Chapter III.

The construction of *regular* continuous strong Markov processes by time change of a Wiener process followed by a space transformation is well-known and goes back to VOLKONSKI [46] as well as ITÔ, MCKEAN [31]. While VOLKONSKI, ITÔ, MCKEAN have used the right-inverse of a perfect additive functional for the time change our approach is based on the notion of a weakly additive functional. With regard to the classification of points of the state interval this enables us to construct an arbitrary continuous strong Markov process.

We mention here that in [31], ch. 5.10, ITÔ, MCKEAN constructed a continuous strong Markov process taking values in $[0,1]$ and admitting regular and right singular points by putting pieces of trajectories together. If moreover the set of right singular points does not contain any interval ITÔ, MCKEAN also indicate a construction by time change (see [31], p. 195).

Now we are going to construct a continuous strong Markov process taking values in an interval $I \subseteq \mathbb{R}$ for arbitrarily given regular points, singular points, scale function and speed measure.

(6.1) Let $R \subseteq I$ be an arbitrary open subset of $\mathbb{R}$ and denote by

$$R = \bigcup_n (a_n, b_n)$$

its decomposition into components. Further let $K_+ \subseteq I$ and $K_- \subseteq I$ be two sets being closed to the right-hand side and left-hand side, respectively, and satisfying

$$I \setminus R = K_+ \cup K_-.$$

We suppose that

$$\sup(I) \in K_- \quad \text{if} \quad \sup(I) \in I$$

and

$$\inf(I) \in K_+ \quad \text{if} \quad \inf(I) \in I.$$

Observe that the required topological properties of K_+, K_- are necessary for K_+, K_- to be a set of right, respectively left singular points (cf. (2.5)(i)). Set

$$\begin{aligned} E &= K_+ \cap K_-, \\ \bar{R} &= \bigcup_n (a_n, b_n) \cup (\{a_n : a_n \in K_+\} \cup \{b_n : b_n \in K_-\}), \\ S_+ &= K_+ \setminus (\{a_n : a_n \in K_+\} \cup E), \\ S_- &= K_- \setminus (\{b_n : b_n \in K_-\} \cup E). \end{aligned}$$

(6.2) For $x \in I$ we define

$$f_-(x) = \begin{cases} b_n & : x \in (a_n, b_n),\ a_n \in K_- \setminus E,\ b_n \in K_- \\ \inf\{y \geq x : y \in K_- \setminus E\} : & \text{otherwise} \end{cases}$$

$$f_+(x) = \begin{cases} a_n & : x \in (a_n, b_n),\ a_n \in K_+,\ b_n \in K_+ \setminus E \\ \sup\{y \leq x : y \in K_+ \setminus E\} : & \text{otherwise} \end{cases}$$

and

$$I(x) = [f_+(x), f_-(x)].$$

It is convenient to introduce the following notations:

$$\begin{aligned} N_{+-} &= \{n : a_n \in K_+,\ b_n \in K_-\},^1 \\ N_{++} &= \{n : a_n \in K_+,\ b_n \in K_+ \setminus E\}, \\ N_{--} &= \{n : a_n \in K_- \setminus E,\ b_n \in K_-\}, \\ N_{-+} &= \{n : a_n \in K_- \setminus E,\ b_n \in K_+ \setminus E\} \end{aligned}$$

and

$$\begin{aligned} I_n &= [a_n, b_n] \cap I \text{ if } n \in N_{+-}, \\ I_n &= [a_n, b_n) \cap I \text{ if } n \in N_{++}, \\ I_n &= (a_n, b_n] \cap I \text{ if } n \in N_{--}, \\ I_n &= (a_n, b_n) \quad\quad \text{if } n \in N_{-+}. \end{aligned}$$

(6.3) Let $p : I \to \mathbb{R}$ denote a continuous strictly increasing function. The restriction of p to R will serve as scale function of the process we are going to construct. The canonical extension of p to the closure $\bar{I} \subseteq \bar{\mathbb{R}}$ of I we also denote by p.

[1] For $a_n \notin I$ (resp. $b_n \notin I$) we make the formal agreement that $a_n \in K_+$ (resp. $b_n \in K_-$).

(6.4) Let m be a measure on $(I, \mathfrak{B}(I))$ possessing the following properties:

(i) $m(G) > 0$ for every open subset $G \subseteq I \setminus E$.

(ii) $m(K) < \infty$ for every compact subinterval $K \subseteq R$.

(iii) $m(\{x\}) = 0$ for every $x \in S_+ \cup S_-$.

(iv) For all $x \in K_+ \setminus E$ resp. $x \in K_- \setminus E$ there exists an $\varepsilon > 0$ such that

$$\sum_n \int_{[a_n,b_n)\cap[x,x+\varepsilon)} (p(b_n \wedge (x+\varepsilon)) - p(y))\, m(dy) + m([x, x+\varepsilon)\cap S_+) < \infty$$

resp.

$$\sum_n \int_{(a_n,b_n]\cap(x-\varepsilon,x]} (p(y) - p(a_n \vee (x-\varepsilon)))\, m(dy) + m((x-\varepsilon, x]\cap S_-) < \infty.$$

(v) For all $x \in I$ satisfying $[x, \infty) \cap K_- = \emptyset$ resp. $(-\infty, x] \cap K_+ = \emptyset$,

$$\int_{I\cap[x,\infty)} \sum_n \mathbf{1}_{[a_n,b_n)}(y)(p(b_n) - p(y))\, m(dy) + m([x,\infty)\cap S_+) = \infty$$

resp.

$$\int_{I\cap(-\infty,x]} \sum_n \mathbf{1}_{(a_n,b_n]}(y)(p(y) - p(a_n))\, m(dy) + m((-\infty, x]\cap S_-) = \infty.$$

(vi) For all $x \in I$ and $z_1 = \inf\{y \geq x : y \in E \cap I(x)\} < \infty$ and $z_2 = \sup\{y \leq x : y \in E \cap I(x)\} > -\infty$ it holds that

$$\int_{U_{z_i}} (r^x(y) + \mathbf{1}_{\bar{R}^c \cap I(x)}(y))\, m(dy) = \infty$$

for all neighbourhoods U_{z_i} of z_i in I, $i = 1, 2$, where

$$\begin{aligned} r^x(y) &= \sum_{n\in N_{+-},\, I_n \subseteq I(x)} \mathbf{1}_{I_n}(y) \\ &+ \sum_{n\in N_{++},\, I_n \subseteq I(x)} \mathbf{1}_{I_n}(y)(p(b_n) - p(y)) \\ &+ \sum_{n\in N_{--},\, I_n \subseteq I(x)} \mathbf{1}_{I_n}(y)(p(y) - p(a_n)) \end{aligned}$$

is defined for all $y \in I$.

We call a measure m on $(I, \mathfrak{B}(I))$ with the properties (6.4)(i)-(vi) a *p-admissible measure.* Observe that, in view of (5.34), the properties (i)-(v) are necessary for $m|_{I \setminus E}$ to be the speed measure associated with the scale p of a continuous strong Markov process.

To begin with we first investigate the case $I = \mathbb{R}$ and $p(x) = x$. The general case will be reduced to this one, later.

(6.5) Theorem *Let $I = \mathbb{R}$, $p(x) = x$, $x \in I$, and let m denote a p-admissible measure. For $A \in \mathfrak{B}(I)$ we set*

$$\textbf{(6.6)} \qquad \tilde{m}(A) = m(A \setminus (\{a_n : a_n \in K_+\} \cup \{b_n : b_n \in K_-\})) + \frac{1}{2} \sum_{a_n \in K_+} m(\{a_n\} \cap A) + \frac{1}{2} \sum_{b_n \in K_-} m(\{b_n\} \cap A).$$

Let be $(W, \mathbb{G})$ a Wiener process on $(\Omega, \mathcal{F}, \mathbf{P}_x, x \in \mathbb{R})$ according to (1.11) with shift operators $\Theta = (\theta_t)_{t \geq 0}$. We define

$$R^* = \bigcup_{n \in N_{-+}} I_n, \quad \tau = \inf\{s \geq 0 : W_s \notin R^*\},$$

$$\textbf{(6.7)} \quad A_t = \begin{cases} \int_{R^*} L^W(t \wedge \tau, y)\, \tilde{m}(dy) \\ + \sum_{n \in N_{+-}} \mathbf{1}_{I_n \subseteq I(W_\tau)} \int_{I_n} L^W(t, y)\, \tilde{m}(dy) \\ + \sum_{n \in N_{++}} \mathbf{1}_{I_n \subseteq I(W_\tau)} \int_{I_n} L^W(t \wedge \mathrm{D}_{b_n}(W), y)\, \tilde{m}(dy) \\ + \sum_{n \in N_{--}} \mathbf{1}_{I_n \subseteq I(W_\tau)} \int_{I_n} L^W(t \wedge \mathrm{D}_{a_n}(W), y)\, \tilde{m}(dy) \\ + \tilde{m}\left([\min_{s \leq t} W_s, \max_{s \leq t} W_s] \cap I(W_\tau) \cap \bar{R}^c\right), \ t \geq 0, \end{cases}$$

and

$$T_t = \inf\{s \geq 0 : A_{s+} > t\}, \qquad t \geq 0.$$

Then $(X, \mathbb{F}) = (W \circ T, \mathbb{F}^W \circ T)$ is a continuous strong Markov process on $(\Omega, \mathcal{F}, \mathbf{P}_x, x \in I)$ with shift operators $\Theta \circ T$ whose set of regular, left singular and right singular points is R, K_- and K_+, respectively. Further, the function p is a scale function of X and $m|_{I \setminus R}$ is just the speed measure of X associated with p.

Before we start proving the theorem let us clarify the definition (6.7) of A and, in particular, the role played by the stopping time τ and the interval $I(W_\tau)$.

Let $(X, \mathbb{F})$ on $(\Omega, \mathcal{F}, \mathbf{P}_x, x \in I)$ be a continuous strong Markov process admitting R, K_+, K_- as the set of regular, right singular, and left singular points, respectively. Clearly, for $x \in I$, $X \in I(x)$ holds $\mathbf{P}_x$-a.s. For $x \in R^*$ the behaviour of X is particularly involved. For example, in the case of $x \in I_n = (a_n, b_n) \subseteq R^*$ we have $X_t \in [f_+(a_n), a_n]$, $t \geq \mathrm{D}_{a_n}(X)$, on $\{\mathrm{D}_{(R^*)^c}(X) = \mathrm{D}_{a_n}(X) < \infty\}$ but $X_t \in [b_n, f_-(b_n)]$, $t \geq \mathrm{D}_{b_n}(X)$, on

$\{\mathrm{D}_{(R^*)^c}(X) = \mathrm{D}_{b_n}(X) < \infty\}$ $\mathbf{P}_x$-a.s. All in all, the process X "lives" in $I(X_{\mathrm{D}_{(R^*)^c}(X)})$ on $[\mathrm{D}_{(R^*)^c}(X), \infty)$ $\mathbf{P}_x$-a.s., $x \in I$.

In our definition (6.7) the role of the stopping time τ and the interval $I(W_\tau)$ is just to take account of that property of X. To be more precise, for $x \notin R^*$ it follows that $\tau = 0$, $I(W_\tau) = I(x)$ $\mathbf{P}_x$-a.s. and the process A defined in (6.7) only "increases" in time points where $W \in I(x)$. Therefore, under $\mathbf{P}_x$ the time changed process $W \circ T$ only takes values in $I(x)$. But if $x \in R^*$ then A_t "increases" on $[\tau, \infty)$ at most in moments t where $W_t \in I(W_\tau)$. For the time changed process $X = W \circ T$ this implies $X \in I(X_{\mathrm{D}_{(R^*)^c}(X)})$ on $[\mathrm{D}_{(R^*)^c}(X), \infty)$ $\mathbf{P}_x$-a.s. in that case, too.

(6.8) Remark In view of the preceding comments it seems to be more natural to define

$$I(x) = [\sup\{y \leq x : y \in K_+\}, \inf\{y \geq x : y \in K_-\}], \qquad x \in I,$$

instead of our definition in (6.2). However, for $x \in E$ this would imply $I(x) = \{x\}$ and even if we modify (6.7) accordingly, as a consequence, we would have to set $m(\{x\}) = \infty$, $x \in E$, in place of (6.4)(vi) to ensure that each $x \in E$ is absorbing for $W \circ T$. On the other hand, from $m(\{x\}) = \infty$, $x \in E$, it follows that (6.4)(vi) is fulfilled. Since we attempt to require as few as possible restrictions, we feel that our definition is more appropriate.

Proof of Theorem (6.5). Since the proof of the theorem is rather involved we decompose it into parts formulated and proved as lemmata.

First observe that $I(W_\tau)$ and thus A are well-defined since τ is finite. Our first goal is to prove that $A^+ = (A_{t+})_{t \geq 0}$ is a weakly additive functional of $(W, \mathbb{G})$ satisfying the conditions (i)-(v) of (3.6).

At first we claim that A is $\mathbb{F}^W$-adapted. Then, using the right continuity of $\mathbb{F}^W$, the same holds true for A^+. We consider the two parts

$$A_t^1 = \int_{R^*} L^W(t \wedge \tau, y)\, \tilde{m}(dy)$$

and

$$A_t^2 = \begin{cases} \sum_{n \in N_{+-}} \mathbf{1}_{I_n \subseteq I(W_\tau)} \int_{I_n} L^W(t, y)\, \tilde{m}(dy) \\ + \sum_{n \in N_{++}} \mathbf{1}_{I_n \subseteq I(W_\tau)} \int_{I_n} L^W(t \wedge \mathrm{D}_{b_n}(W), y)\, \tilde{m}(dy) \\ + \sum_{n \in N_{--}} \mathbf{1}_{I_n \subseteq I(W_\tau)} \int_{I_n} L^W(t \wedge \mathrm{D}_{a_n}(W), y)\, \tilde{m}(dy) \\ + \tilde{m}\left([\min_{s \leq t} W_s, \max_{s \leq t} W_s] \cap I(W_\tau) \cap \bar{R}^c\right) \end{cases}$$

of A. Since τ is an $\mathbb{F}^W$-stopping time, A^1 is $\mathbb{F}^W$-adapted. For $t < \tau$ we have $A_t^2 = 0$ (see also (1.17)(i)) and A^2 is $\mathbb{F}^W$-adapted, too.

(6.9) Lemma (i) *Let $x \notin R^*$ and define*

$$A_t^x = \begin{cases} \sum_{n \in N_{+-}} \mathbf{1}_{I_n \subseteq I(x)} \int_{I_n} L^W(t,y)\,\tilde{m}(\mathrm{d}y) \\ + \sum_{n \in N_{++}} \mathbf{1}_{I_n \subseteq I(x)} \int_{I_n} L^W(t \wedge \mathrm{D}_{b_n}(W),y)\,\tilde{m}(\mathrm{d}y) \\ + \sum_{n \in N_{--}} \mathbf{1}_{I_n \subseteq I(x)} \int_{I_n} L^W(t \wedge \mathrm{D}_{a_n}(W),y)\,\tilde{m}(\mathrm{d}y) \\ + \tilde{m}\left([\min_{s \le t} W_s, \max_{s \le t} W_s] \cap I(x) \cap \bar{R}^c\right),\ t \ge 0. \end{cases}$$

Then for $z = \inf\{y \ge x : y \in E \cap I(x)\} < \infty$ or $z = \sup\{y \le x : y \in E \cap I(x)\} > -\infty$ we have

$$\mathbf{P}_z(\{A_t^x = \infty\}) = 1, \qquad t > 0.$$

(ii) *Let*

$$\xi = \begin{cases} \mathrm{D}_{a_m} \mathbf{1}_{\{\mathrm{D}_{a_m} < \mathrm{D}_{b_m}\}} + \mathrm{D}_{z_m} \mathbf{1}_{\{\mathrm{D}_{a_m} \ge \mathrm{D}_{b_m}\}} & : \quad W_\tau \in (a_m, b_m),\ m \in N_{++},\ a_m \in E \\ \mathrm{D}_{b_m} \mathbf{1}_{\{\mathrm{D}_{b_m} < \mathrm{D}_{a_m}\}} + \mathrm{D}_{z_m} \mathbf{1}_{\{\mathrm{D}_{a_m} \le \mathrm{D}_{b_m}\}} & : \quad W_\tau \in (a_m, b_m),\ m \in N_{--},\ b_m \in E \\ \inf\{s \ge \tau : W_s \in E \cap I(W_\tau)\} & : \quad \textit{otherwise} \end{cases}$$

where

$$z_m = \begin{cases} \inf\{y > a_m : y \in E \cap I(a_m)\} & : \quad m \in N_{++} \\ \sup\{y < b_m : y \in E \cap I(b_m)\} & : \quad m \in N_{--} \end{cases}.$$

Then,

$$T_\infty \le \xi$$

holds a.s.

Proof of (6.9). (i) For z finite (6.4)(vi) implies

$$\int_{U_z} (r^x(y) + \mathbf{1}_{\bar{R}^c \cap I(x)}(y))\,\tilde{m}(\mathrm{d}y) = \infty$$

for all neighbourhoods U_z of z since $\tilde{m} \ge \frac{1}{2}m$. In the case of

$$\int_{U_z} \mathbf{1}_{\bar{R}^c \cap I(x)}(y)\,\tilde{m}(\mathrm{d}y) = \infty$$

for all neighbourhoods U_z of z the assertion of (i) follows, immediately. Now suppose

$$\int_{(z-\varepsilon, z+\varepsilon)} r^x(y)\,\tilde{m}(\mathrm{d}y) = \infty$$

for all $\varepsilon > 0$. Then $\mathbf{P}_z$-a.s.

$$\begin{aligned}2A^x_{\mathrm{D}_{z-\varepsilon}\vee \mathrm{D}_{z+\varepsilon}} \geq & \textstyle\sum_{n\in N_{+-}} \mathbf{1}_{I_n\subseteq I(x)} \int_{I_n\cap[z,z+\varepsilon)} L^W(\mathrm{D}_{z+\varepsilon},y)\,\widetilde{m}(\mathrm{d}y) + \\ & + \textstyle\sum_{n\in N_{++}} \mathbf{1}_{I_n\subseteq I(x)} \int_{I_n\cap[z,z+\varepsilon)} L^W(\mathrm{D}_{z+\varepsilon}\wedge \mathrm{D}_{b_n}(W),y)\,\widetilde{m}(\mathrm{d}y) \\ & + \textstyle\sum_{n\in N_{+-}} \mathbf{1}_{I_n\subseteq I(x)} \int_{I_n\cap(z-\varepsilon,z]} L^W(\mathrm{D}_{z-\varepsilon},y)\,\widetilde{m}(\mathrm{d}y) \\ & + \textstyle\sum_{n\in N_{--}} \mathbf{1}_{I_n\subseteq I(x)} \int_{I_n\cap(z-\varepsilon,z]} L^W(\mathrm{D}_{z-\varepsilon}\wedge \mathrm{D}_{a_n}(W),y)\,\widetilde{m}(\mathrm{d}y).\end{aligned}$$

Applying Proposition (A1.8) we obtain

$$A^x_{\mathrm{D}_{z-\varepsilon}\vee \mathrm{D}_{z+\varepsilon}} = \infty \qquad \mathbf{P}_z - \text{a.s.}$$

for all $\varepsilon > 0$ proving (i) in the remaining case, too.

(ii) The assertion is equivalent to

$$A_{\xi+t} = \infty \ \mathbf{P}_x - \text{a.s.} \quad \text{on} \quad \{\xi < \infty\} \tag{6.10}$$

for all $t > 0$ and $x \in I$. Let us fix $x \in I$.

For $x \in E$ we have $\tau = 0$, $W_\tau = x$ and $\xi = 0$ $\mathbf{P}_x$-a.s. Furthermore from (i) follows $A_t = \infty$ $\mathbf{P}_x$-a.s., $t > 0$, proving (6.10) in that case.

Now, we assume $x \notin E$. Then, the definition of τ yields

$$\mathbf{P}_x(\{W_\tau \notin E\}) = 1. \tag{6.11}$$

Obviously we have $\xi \geq \tau$. To prove (6.10) we want to apply the strong Markov property of W in τ. For this purpose we first verify that

$$A_t \circ \theta_\tau + A_\tau = A_{t+\tau} \tag{6.12}$$

for all $t \geq 0$ $\mathbf{P}_x$-a.s. Let us show this for each summand in (6.7). It is easy to see that $\tau \circ \theta_\tau = 0$, $I(W_\tau) \circ \theta_\tau = I(W_\tau)$ holds $\mathbf{P}_x$-a.s. On the other hand, (1.16)(ii) yields

$$L^W(t\wedge\eta,y)\circ\theta_\tau + L^W(\tau\wedge\eta,y) = L^W((t+\tau)\wedge\eta,y), \quad t \geq 0,\, y \in \bar{R}, \text{ a.s.,}$$

where η stands for τ, D_{a_n}, D_{b_n} or ∞. This proves (6.12) for every summand in (6.7) containing the local time of W. Finally, using (6.11) and (6.4)(iii) we get

$$\begin{aligned} & \widetilde{m}\left(\left[\min_{s\leq t} W_s, \max_{s\leq t} W_s\right] \cap I(W_\tau)\cap \bar{R}^c\right)\circ\theta_\tau \\ + \ & \widetilde{m}\left(\left[\min_{s\leq \tau} W_s, \max_{s\leq \tau} W_s\right] \cap I(W_\tau)\cap \bar{R}^c\right) \\ = \ & \widetilde{m}\left(\left[\min_{s\leq t+\tau} W_s, \max_{s\leq t+\tau} W_s\right] \cap I(W_\tau)\cap \bar{R}^c\right) \end{aligned} \tag{6.13}$$

for all $t \geq 0$ $\mathbf{P}_x$-a.s. proving (6.12).

Proceeding with the proof of (6.10) we remark that $\xi \circ \theta_\tau = \xi - \tau$ is an easy consequence of the definition of τ. Now we combine (6.11),(6.12) and the strong Markov property of W to obtain

$$\begin{aligned}
\mathbf{P}_x(\{A_{\xi+t}=\infty,\,\xi<\infty\}) &= \mathbf{P}_x(\{A_{\xi+t}=\infty,\,A_\tau=\infty,\,\xi<\infty\})\\
&\quad+\mathbf{P}_x(\{A_{\xi+t}=\infty,\,A_\tau<\infty,\,\xi<\infty\})\\
&= \mathbf{P}_x(\{A_{\xi+t}=\infty,\,A_\tau=\infty,\,\xi<\infty\})\\
&\quad+\mathbf{P}_x(\{A_{\xi+t}\circ\theta_\tau=\infty,\,A_\tau<\infty,\,\xi\circ\theta_\tau<\infty\})\\
&\geq \mathbf{P}_x(\{A_\tau=\infty,\,\xi<\infty\})\\
&\quad+\mathbf{E}_x\mathbf{P}_{W_\tau}(\{A_{\xi+t}=\infty,\,\xi<\infty\})\mathbf{1}_{\{A_\tau<\infty\}}.
\end{aligned}$$

Therefore, in view of $W_\tau \notin R^*$, it suffices to show (6.10) for $x \notin R^* \cup E$, only. We set

$$z_u = \sup\{y \leq x : y \in E \cap I(x)\},$$
$$z_o = \inf\{y \geq x : y \in E \cap I(x)\}.$$

Clearly, if z_i is finite then combining (6.1) and the definition of $I(x)$ we see that $z_i \in E \cap I(x)$, $i = u, o$.

In what follows we distinguish several cases concerning the position of our point $x \notin R^* \cup E$.

First let $x \in S_+$ or $x \in I_m$, $m \in N_{++} \cup N_{+-}$ with $a_m \in K_+ \setminus E$. Then, $z_u = -\infty$ and $\xi = \mathrm{D}_{z_o}$. In the case of $\mathbf{P}_x(\{\xi < \infty\}) > 0$ we have $z_o < \infty$. Using (1.16) and (1.17)(ii) we consequently obtain the estimate

$$\begin{aligned}
A_{\xi+t} = A_{\mathrm{D}_{z_o}+t} &\geq \sum_{n\in N_{+-}} \mathbf{1}_{I_n\subseteq I(x)} \int_{I_n} L^W(\mathrm{D}_{z_o}+t, y)\,\tilde{m}(dy)\\
&+ \sum_{n\in N_{++}} \mathbf{1}_{I_n\subseteq I(x)} \int_{I_n} L^W((\mathrm{D}_{z_o}+t)\wedge \mathrm{D}_{b_n}(W), y)\,\tilde{m}(dy)\\
&+ \tilde{m}\left(\left[\min_{\mathrm{D}_{z_o}\leq s\leq t+\mathrm{D}_{z_o}} W_s, \max_{\mathrm{D}_{z_o}\leq s\leq t+\mathrm{D}_{z_o}} W_s\right]\cap I(x)\cap \bar{R}^c\right)\\
&\geq A_t^x \circ \theta_{\mathrm{D}_{z_o}}
\end{aligned} \tag{6.14}$$

on $\{\xi < \infty\}$ for all $t > 0$ $\mathbf{P}_x$-a.s. where A^x is defined as in (6.9)(i). Using (6.9)(i) and the strong Markov property of W this leads to (6.10).

Analogously, one can treat the case $x \in S_-$ or $x \in I_m$, $m \in N_{--} \cup N_{+-}$ with $b_m \in K_- \setminus E$.

Next let $x \in I_m$, $m \in N_{+-}$ with $a_m \in E$ and $b_m \in E$. Then $z_u = a_m$, $z_o = b_m$ and $\xi = \mathrm{D}_{z_u} \wedge \mathrm{D}_{z_o}$ $\mathbf{P}_x$-a.s. As above on $\{\mathrm{D}_{z_u} < \mathrm{D}_{z_o}\}$ we get

$$A_{\xi+t} = A_{\mathrm{D}_{z_u}+t} \geq A_t^x \circ \theta_{\mathrm{D}_{z_u}}, \qquad t \geq 0,\ \mathbf{P}_x - \text{a.s.},$$

and an analogous result is true on $\{\mathrm{D}_{z_o} < \mathrm{D}_{z_u}\}$. Applying (6.9)(i) again, this yields (6.10).

Finally, we treat the case $x \in (a_m, b_m)$, $m \in N_{++}$ with $a_m \in E$, the proof for the case $x \in (a_m, b_m)$, $m \in N_{--}$ with $b_m \in E$, is analogously. Here, $I(x) = [a_m, f_-(x)]$ and $z_u = a_m$ which leads to $\xi = \mathrm{D}_{z_u}$ on $\{\mathrm{D}_{z_u} < \mathrm{D}_{b_m}\}$. As in (6.14) on $\{\mathrm{D}_{z_u} < \mathrm{D}_{b_m}\}$ we compute

$$A_{\xi+t} = A_{\mathrm{D}_{z_u}+t} \geq A_t^x \circ \theta_{\mathrm{D}_{z_u}}, \qquad t \geq 0, \mathbf{P}_x - \text{a.s.},$$

implying (6.10) on this set. If $\mathbf{P}_x(\{\xi < \infty\} \cap \{\mathrm{D}_{z_u} \geq \mathrm{D}_{b_m}\}) > 0$ we have $z_o < \infty$ and the inequality (6.14) follows on $\{\xi < \infty\} \cap \{\mathrm{D}_{z_u} \geq \mathrm{D}_{b_m}\}$. Applying (6.9)(i) once more, the proof of (6.10) is complete. □

(6.15) Lemma *The equality $A_\infty^+ = A_\infty = \infty$ holds a.s.*

Proof of (6.15). Let $x \in I$ be fixed. If $x \in E$ then the assertion immediately follows from (6.9)(i) and so we assume $x \notin E$ in what follows. Using (6.12) we obtain

$$\mathbf{P}_x(\{A_\infty = \infty\}) \geq \mathbf{P}_x(\{A_\infty \circ \theta_\tau = \infty\}) = \mathbf{E}_x \mathbf{P}_{W_\tau}(\{A_\infty = \infty\});$$

therefore, it suffices to prove the assertion with respect to $\mathbf{P}_x$ for $x \notin R^* \cup E$.

For $x \notin R^*$ we have $\tau = 0$, $I(W_\tau) = I(x)$ $\mathbf{P}_x$-a.s.

If there exists an $n \in N_{+-}$ satisfying $I_n \subseteq I(x)$ then (6.15) is an easy consequence of (6.4)(i) and (A1.9)(i).

In case of $\mathbf{P}_x(\{\xi < \infty\}) = 1$ with ξ from (6.9)(ii) the assertion follows from (6.9)(ii). Now assume $\mathbf{P}_x(\{A_\infty < \infty\} \cap \{\xi = \infty\}) > 0$ then $I_n \not\subseteq I(x)$ for all $n \in N_{+-}$ and at least one of the two cases $[x, \infty) \cap I \cap K_- = \emptyset$ or $(-\infty, x] \cap I \cap K_+ = \emptyset$ must be satisfied. Without loss of generality we only consider the first case. Then we have $f_-(x) = \infty$ leading to

$$A_\infty \geq \sum_{n \in N_{++}} \int_{I_n \cap [x,\infty)} L^W(\mathrm{D}_{b_n}, y)\, \tilde{m}(dy) + \tilde{m}([x, \infty) \cap S_+)$$

$\mathbf{P}_x$-a.s. Combining (6.4)(v) and (A1.8) this yields $A_\infty = \infty$ $\mathbf{P}_x$-a.s. □

(6.16) Lemma *A^+ is continuous on $[0, T_\infty)$.*

Proof of (6.16). If $t < T_\infty$ then $A_t^+ < \infty$ and, therefore, the summands in (6.7) involving the local time of W are continuous on $[0, T_\infty)$. Further, in the case of $t < T_\infty$ we have $[\min_{s \leq t} W_s, \max_{s \leq t} W_s] \cap I(W_\tau) \cap E = \emptyset$ because (6.9)(ii) implies $A_t^+ = \infty$, otherwise. Hence, the remaining summand

$$\tilde{m}([\min_{s \leq t} W_s, \max_{s \leq t} W_s] \cap I(W_\tau) \cap \bar{R}^c)$$

is continuous on $[0, T_\infty)$ by (6.4)(iii). □

(6.17) Lemma *It holds that*

$$\mathbf{P}_x(\{A_{0+} < \infty\}) = 1 \quad \textit{if} \quad x \notin E$$

and

$$\mathbf{P}_x(\{A_{0+} = \infty\}) = 1 \quad \textit{if} \quad x \in E.$$

Proof of (6.17). First let $x \in R$. Then there exists an $\varepsilon > 0$ such that $U_x = [x-\varepsilon, x+\varepsilon] \subseteq R$ because R is open. Thus, we have

$$A_t = \int_{U_x} L^W(t,y)\,\tilde{m}(dy)$$

for all $t \leq D_{U_x^c}(W)$ $\mathbf{P}_x$-a.s. Combining (6.4)(ii) and (A1.9)(ii), this implies

$$\mathbf{P}_x(\{A_t < \infty,\, t \leq D_{U_x^c}\}) = 1.$$

Now let $x \in K_+ \setminus E$. Then there exists an $\varepsilon > 0$ such that $[x, x+\varepsilon) \subseteq R \cup K_+ \setminus E$ because K_- is assumed to be closed to the left-hand side. Now

$$A_{D_{x+\varepsilon}(W)} = \sum_n \int_{[a_n,b_n)\cap[x,x+\varepsilon)} L^W(D_{b_n} \wedge D_{x+\varepsilon}, y)\,\tilde{m}(dy) + \tilde{m}([x,x+\varepsilon)\cap S_+)$$

follows $\mathbf{P}_x$-a.s. Combining (6.4)(iv) and (A1.8), for sufficiently small $\varepsilon > 0$ the last expression is finite.

For $x \in K_- \setminus E$ the proof is analogously while for $x \in E$ the assertion is an easy consequence of (6.9)(i). □

(6.18) Lemma $(A^+, \mathbb{G})$ *is a weakly additive functional of* $(W, \mathbb{G})$.

Proof of (6.18). We have to verify the properties demanded in (3.4).

First, the $\mathcal{F}_\infty^W$-measurability of A_t^+, $t \geq 0$, is obvious.

Next, let (s,ω) be an element of the set $H_+(A^+)$ defined in (3.3). Then, $s < T_\infty(\omega)$ and (6.16) implies $A_s(\omega) = A_s^+(\omega)$. Therefore, we have to verify that

(6.19) $$A_s(\omega) + A_t \circ \theta_s(\omega) = A_{s+t}(\omega)$$

for all $(s,\omega) \in H_+(A^+)$, $t \geq 0$, a.s.

In what follows we fix a measure $\mathbf{P}_x$, $x \in I = \mathbb{R}$. Further, let (s,ω) be an arbitrary element of $H_+(A^+)$ whose component ω lies outside a certain $\mathbf{P}_x$-null set N being independent of s. This exceptional set N will become clear during the proof. Since ω is fixed, simplifying our notation the argument ω will be omitted, hereafter.

We begin by considering the case $s < \tau$. Then, $I(W_\tau) \circ \theta_s = I(W_\tau)$, $s < D_{a_n}$, $s < D_{b_n}$, $\forall n$, holds and, consequently, for $\eta = \tau$, D_{a_n}, D_{b_n}, ∞, (1.16)(ii) guarantees the equality

$$L^W((t+s)\wedge\eta, y) = L^W(t\wedge\eta, y)\circ\theta_s + L^W(s\wedge\eta, y)$$

for all $t \geq 0$ and $y \in \mathbb{R}$. Moreover, $[\min_{u\leq s} W_u, \max_{u\leq s} W_u] \cap \bar{R}^c = \emptyset$ which yields

$$\tilde{m}([\min_{u\leq t} W_u, \max_{u\leq t} W_u]\cap I(W_\tau)\cap \bar{R}^c)\circ\theta_s + \tilde{m}([\min_{u\leq s} W_u, \max_{u\leq s} W_u]\cap I(W_\tau)\cap \bar{R}^c)$$

$$= \tilde{m}([\min_{u\leq s+t} W_u, \max_{u\leq s+t} W_u]\cap I(W_\tau)\cap \bar{R}^c)$$

for all $t \geq 0$. All in all this proves (6.19) for $s < \tau$.

Now let $s \geq \tau$. As explained before (6.8), on $[\tau, \infty)$ the process A^+ only "increases" in time points where $W \in I(W_\tau)$. Thus, the definition of $H_+(A^+)$ and $s \geq \tau$ yield $W_s \in I(W_\tau) \subseteq (R^*)^c$ leading to $\tau \circ \theta_s = 0$. Consequently, by (1.16) we have

$$\int_{R^*} L^W((t+s)\wedge\tau, y)\,\tilde{m}(dy) = \int_{R^*} L^W(s\wedge\tau, y)\,\tilde{m}(dy) + \int_{R^*} L^W(t\wedge\tau, y)\,\tilde{m}(dy)\circ\theta_s$$

for all $t \geq 0$ and (6.19) has already been shown for the first summand in (6.7). The property claimed in (6.19) remains to be checked correspondingly for the remaining summands in (6.7).

Using $W_s \in I(W_\tau)$ and $\tau \circ \theta_s = 0$ again, we deduce

(6.20) $$I(W_\tau)\circ\theta_s = I(W_s) \subseteq I(W_\tau).$$

In view of $\tau \leq s < T_\infty$ and (6.9)(ii) we also have

(6.21) $$W_\tau \notin E.$$

In what follows we will distinguish various cases concerning the position of the point W_τ.

First of all let $W_\tau \in I_{n_0}$, $n_0 \in N_{+-}$. Then the definition of $I(W_\tau)$, $s < T_\infty$ and (6.9)(ii) imply $a_{n_0} \leq W_s \leq b_{n_0}$ and, therefore,

$$I(W_\tau)\circ\theta_s = I(W_s) = I(W_\tau)$$

holds. Now we verify the equality corresponding with (6.19) for each of the remaining summands in (6.7).

If $n \in N_{+-}$ then (1.16) guarantees

$$\int_{I_n} L^W(t+s, y)\,\tilde{m}(dy) = \int_{I_n} L^W(s, y)\,\tilde{m}(dy) + \int_{I_n} L^W(t, y)\,\tilde{m}(dy)\circ\theta_s$$

for all $t \geq 0$.

If $n \in N_{++}$ with $I_n \subseteq I(W_\tau)$ then from the definition of $I(W_\tau)$ and $W_\tau \in I_{n_0}$ it follows that $b_{n_0} \leq a_n$ as well as $b_{n_0} \in E$. Combining $s < T_\infty$ and (6.9)(ii) this gives $s < \mathrm{D}_{b_{n_0}}$ and thus also $s < \mathrm{D}_{b_n}$. Applying (1.16)(ii) again, we obtain

$$\int_{I_n} L^W((t+s)\wedge \mathrm{D}_{b_n}, y)\,\tilde{m}(dy) = \int_{I_n} L^W(s\wedge \mathrm{D}_{b_n}, y)\,\tilde{m}(dy) + \int_{I_n} L^W(t\wedge \mathrm{D}_{b_n}, y)\,\tilde{m}(dy)\circ\theta_s$$

for all $t \geq 0$.

The case $n \in N_{--}$, $I_n \subseteq I(W_\tau)$, can be treated analogously.

To prove (6.19) for the last summand in (6.7) we use the definition of $I(W_\tau)$, $s < T_\infty$ and (6.9)(ii) to derive that

$$[\min_{u\leq s} W_u, \max_{u\leq s} W_u] \cap I(W_\tau) \subseteq I_{n_0} \subseteq \bar{R}.$$

This leads to

$$\tilde{m}([\min_{u\leq t} W_u, \max_{u\leq t} W_u] \cap I(W_\tau) \cap \bar{R}^c) \circ \theta_s + \tilde{m}([\min_{u\leq s} W_u, \max_{u\leq s} W_u] \cap I(W_\tau) \cap \bar{R}^c)$$

$$= \tilde{m}([\min_{u\leq s+t} W_u, \max_{u\leq s+t} W_u] \cap I(W_\tau) \cap \bar{R}^c)$$

for all $t \geq 0$ and, summarizing, (6.19) has been proved for $W_\tau \in I_{n_0}$, $n_0 \in N_{+-}$.

We proceed with the case $W_\tau \in I_{n_0}$, $n_0 \in N_{++}$, or $W_\tau \in S_+$. As above, we investigate each of the remaining summands in (6.7) separately.

For all $n \in N_{--}$, by (6.21) it holds that $I_n \not\subseteq I(W_\tau)$. Therefore, (6.20) implies $I_n \not\subseteq I(W_\tau) \circ \theta_s$, $n \in N_{--}$, and (6.19) is evident for the summands in (6.7) corresponding with $n \in N_{--}$.

If $n \in N_{+-}$ with $I_n \subseteq I(W_\tau)$ then the definition of $I(W_\tau)$ supplies $W_\tau \leq a_n$. Moreover, using $W_s \in I(W_\tau)$, $s < T_\infty$ and (6.9)(ii) we have $W_s \leq b_n$ and $I_n \subseteq I(W_s) = I(W_\tau) \circ \theta_s$ follows (see (6.20)). Applying (1.16) we obtain (6.19) for all summands in (6.7) corresponding with $n \in N_{+-}$.

If $n \in N_{++}$ with $I_n \subseteq I(W_\tau)$ then we have $W_\tau \leq a_n$. For $s < \mathrm{D}_{b_n}$, that is $W_s < b_n$, we deduce $I_n \subseteq I(W_s) = I(W_\tau) \circ \theta_s$ leading to

$$\mathbf{1}_{I_n \subseteq I(W_\tau)} \int_{I_n} L^W((t+s) \wedge \mathrm{D}_{b_n}, y)\, \tilde{m}(dy) =$$

$$= \mathbf{1}_{I_n \subseteq I(W_\tau)} \int_{I_n} L^W(s \wedge \mathrm{D}_{b_n}, y)\, \tilde{m}(dy)$$

$$+ \left(\mathbf{1}_{I_n \subseteq I(W_\tau)} \int_{I_n} L^W(t \wedge \mathrm{D}_{b_n}, y)\, \tilde{m}(dy) \right) \circ \theta_s$$

for all $t \geq 0$ by (1.16)(ii). In the case of $s \geq \mathrm{D}_{b_n}$ the process A^+ does not "increase" in s if $W_s < b_n$ and we obtain $W_s \geq b_n$ since $s \in H_+(A^+)$. But this means $I_n \not\subseteq I(W_s) = I(W_\tau) \circ \theta_s$ and the above equality already shown for $s < \mathrm{D}_{b_n}$ is trivially satisfied for $s \geq \mathrm{D}_{b_n}$.

Finally,

$$[\min_{u\leq t} W_u, \max_{u\leq t} W_u] \cap I(W_\tau) \cap \bar{R}^c = [W_\tau, \max_{u\leq t} W_u] \cap I(W_\tau) \cap \bar{R}^c, \quad t \geq 0,$$

and

$$[W_s, y] \cap I(W_\tau) = [W_s, y] \cap I(W_s), \qquad y \in \mathbb{R},$$

hold. Therefore, we obtain

$$\tilde{m}([W_\tau, \max_{u\le t} W_u] \cap I(W_\tau) \cap \bar{R}^c) \circ \theta_s + \tilde{m}([W_\tau, \max_{u\le s} W_u] \cap I(W_\tau) \cap \bar{R}^c)$$

$$= \tilde{m}([W_s, \max_{s\le u\le s+t} W_u] \cap I(W_s) \cap \bar{R}^c) + \tilde{m}([W_\tau, \max_{u\le s} W_u] \cap I(W_\tau) \cap \bar{R}^c)$$

$$= \tilde{m}([W_\tau, \max_{u\le s+t} W_u] \cap I(W_\tau) \cap \bar{R}^c)$$

for all $t \ge 0$ where we have used $[W_s, \max_{u\le s} W_u] \cap \bar{R}^c = \emptyset$ for the last equality.

All in all, (6.19) has also been shown if $W_\tau \in I_{n_0}$, $n_0 \in N_{++}$, or $W_\tau \in S_+$.

Finally, by symmetry, the last case

$$W_\tau \in I_{n_0},\ n_0 \in N_{--} \quad \text{or} \quad W_\tau \in S_-$$

can be treated analogously. This proves that $(A^+, \mathbb{G})$ is a weakly additive functional of $(W, \mathbb{G})$. □

(6.22) Lemma *It holds that*

$$A_t^+ \circ \theta_{T_\infty} = \infty \quad on \quad \bigcup_n \{T_n = T_\infty < \infty\}, \qquad t \ge 0,\ a.s.$$

Proof of (6.22). It suffices to show that

$$\textbf{(6.23)} \qquad T_\infty = \xi \quad \text{on} \quad \bigcup_n \{T_n = T_\infty < \infty\} \text{ a.s.}$$

for the random time ξ defined in (6.9)(ii). Indeed, on $\{T_\infty = \xi < \infty\}$ we have $W_{T_\infty} \in E$ and the assertion follows from (6.17) simply using the strong Markov property of W.

Now let us verify (6.23). (6.9)(ii) just states $T_\infty \le \xi$ a.s. Denote

$$B = \bigcup_n \{T_n = T_\infty < \infty\} \cap \{T_\infty < \xi\}.$$

(6.23) is verified if we can show that $\mathbf{P}_x(B) = 0$, $x \in I$.

First of all, from the definition of B it follows that $A^+_{T_\infty -} < \infty$, $A^+_{T_\infty} = \infty$ on B. Further, (6.9)(ii) implies

$$[\min_{u\le T_\infty} W_u, \max_{u\le T_\infty} W_u] \cap I(W_\tau) \cap \bar{R}^c \cap E = \emptyset \quad \text{on} \quad B \qquad \text{a.s.}$$

Using (6.4)(iii), this yields that the process A is left continuous in T_∞ on B:

$$\infty > A^+_{T_\infty -} = A_{T_\infty -} = \lim_{t\uparrow T_\infty} A_t = A_{T_\infty} \quad \text{on} \quad B.$$

Because of $T_\infty \ge \tau$ a.s. this implies $W_{T_\infty} \in I(W_\tau)$ on B a.s. Using this and taking into account that $T_\infty < \xi$ on B we can repeat the second part of the proof of (6.19) with respect to $(T_\infty(\omega), \omega)$ for almost all $\omega \in B$ although $(T_\infty(\omega), \omega) \notin H_+(A^+)$.

Hence,

$$A_{T_\infty+t} = A_{T_\infty} + A_t \circ \theta_{T_\infty} \quad \text{on} \quad B, \qquad t \geq 0, \text{a.s.}$$

As a consequence, we obtain

(6.24) $$\infty = A^+_{T_\infty} = A_{T_\infty} + A_{0+} \circ \theta_{T_\infty} \quad \text{on} \quad B \text{ a.s.}$$

On B, A_{T_∞} is a.s. finite. We also claim that $A_{0+} \circ \theta_{T_\infty} < \infty$ a.s. on B. In fact, $T_\infty < \xi$ and $W_{T_\infty} \in I(W_\tau)$ on B a.s. imply $W_{T_\infty} \notin E$ on B a.s. and applying (6.17) as well as the strong Markov property of W we have $A_{0+} \circ \theta_{T_\infty} < \infty$ on B a.s. In view of (6.24), $\mathbf{P}_x(B) = 0$, $x \in I$, is proved. □

(6.25) Lemma *The following equality holds:*

$$W_{T_{t-}} = W_{T_t}, \qquad t \geq 0, \; a.s.$$

Proof of (6.25). Using Theorem IV.13 from [12], it suffices to verify that $W_{T_{\eta-}} = W_{T_\eta}$ a.s. for every finite $\mathbb{F}^W \circ T$-stopping time η.

Clearly, on $\{T_{\eta-} = T_\eta\}$ there is nothing to prove. Let B denote the set $\{T_{\eta-} < T_\eta\}$. Then we have

(6.26) $$A^+_{T_{\eta-}} = A^+_t \quad \text{for} \quad t \in [T_{\eta-}, T_\eta) \quad \text{on} \quad B$$

and, as a consequence,

$$T_{\eta-} \geq \tau \quad \text{on} \quad B \qquad \text{a.s.}$$

since A is strictly increasing on $[0, \tau)$ by (6.4)(i),(ii) and (A1.10).

Next we show that

(6.27) $$T_\eta < T_\infty \text{ a.s.} \quad \text{on} \quad B.$$

Indeed, since η is finite, (6.23) gives $T_\eta = \xi$ a.s. on $\{T_\eta = T_\infty\} \cap B$ where ξ is defined in (6.9)(ii). Obviously, T_η is strictly positive on B. In view of (6.9)(ii), the same is true for ξ on B a.s. As a consequence, $W_\tau \notin E$ on B a.s. But from the definition of ξ, (6.4)(i) and (A1.10), $A_{(\xi-\varepsilon)\vee 0} < A_\xi$ on $\{\xi > 0\}$ a.s., $\varepsilon > 0$. Summing up, (6.27) is a consequence of (6.26).

We know from (6.16) that A^+ is continuous on $[0, T_\infty)$. Using (6.27) and that T is the right-inverse of A^+,

(6.28) $$A_{(T_{\eta-}-\varepsilon)\vee 0} < A_{T_{\eta-}}, \; A_{T_\eta} < A_{T_\eta+\varepsilon} \quad \text{on} \quad B$$

for all $\varepsilon > 0$ a.s. Moreover, using $T_{\eta-} \geq \tau$ on B a.s. from this we obtain

$$W_{T_{\eta-}}, W_{T_\eta} \in I(W_\tau) \quad \text{on} \quad B \qquad \text{a.s.}$$

Next we investigate the location of $W_{T_{\eta-}}$. First we assert that

(6.29) $$W_{T_{\eta-}} \notin E \quad \text{on} \quad B \text{ a.s.}$$

In fact, in view of $W_{T_{\eta-}} \in I(W_\tau)$ on B a.s. (6.9)(ii) implies $T_{\eta-} = T_\eta = \xi$ on $\{W_{T_{\eta-}} \in E\} \cap B$ a.s. By the definition of B, this leads to $\mathbf{P}_x(\{W_{T_{\eta-}} \in E\} \cap B) = 0$, $x \in I$.

Now also

(6.30) $$W_{T_\eta -} \notin R \quad \text{on} \quad B \text{ a.s.}$$

Indeed, on $\{W_{T_\eta -} \in R\}$ the stopping time $\zeta = \inf\{s \geq T_{\eta -} : W_s \notin R\}$ satisfies $\zeta > T_{\eta -}$ a.s. and, combining $W_{T_\eta -} \in I(W_\tau)$ on B a.s. as well as (6.9)(i),(ii) and (A1.10), A_t is strictly increasing for all $t \in [T_\eta -, \zeta)$ on $B \cap \{W_{T_\eta -} \in R\}$ a.s. Now (6.26) yields (6.30).

Taking into account (6.29) and (6.30), for the proof of the desired equality

$$W_{T_\eta -} = W_{T_\eta} \quad \text{on} \quad B \qquad \text{a.s.}$$

we only have to deal with the set

$$\{W_{T_\eta -} \in \{a_n : a_n \in K_+ \setminus E\} \cup S_+ \cup \{b_n : b_n \in K_- \setminus E\} \cup S_-\} \cap B.$$

We begin assuming $W_{T_\eta -} = a_n \in K_+ \setminus E$ and our goal is to show that $W_{T_\eta} = a_n$. From the definition of $I(W_\tau)$ it follows that $b_n \in I(W_\tau)$ since $W_{T_\eta -} \in I(W_\tau)$. Moreover, (6.28) implies $T_\eta - < \mathrm{D}_{b_n}$ if $b_n \in K_+ \setminus E$.

Let $W_{T_\eta} > a_n$; then (6.4)(i),(ii) and (A1.10) yield a contradiction of $A_{T_\eta -} = A_{T_\eta}$.

Now let $W_{T_\eta} < a_n$. In the case of $W_\tau \geq a_n$ we get $f_+(W_\tau) \geq a_n$ and $W_{T_\eta} < a_n$ contradicts the fact that $W_{T_\eta} \in I(W_\tau)$. Otherwise, if $W_\tau < a_n$, by combining $a_n = W_{T_\eta -} \in I(W_\tau)$ and (6.9)(ii), we have $[W_\tau, a_n] \subseteq R \cup (K_+ \setminus E)$. Therefore, using the definition of A, $W_{T_\eta} < a_n$ is either inconsistent with the right-hand inequality of (6.28) or with (6.26).

Summarizing, for $W_{T_\eta -} = a_n \in K_+ \setminus E$ we have shown that $W_{T_\eta -} = W_{T_\eta}$.

Next, let $W_{T_\eta -}$ be an element of S_+. Applying (2.11) this implies that $(W_{T_\eta -}, W_{T_\eta -} + \varepsilon) \cap (K_+ \setminus E) \neq \emptyset$, $\varepsilon > 0$. From the definition of $I(W_\tau)$ and $W_{T_\eta -} \in I(W_\tau)$ we deduce $W_\tau \leq W_{T_\eta -}$. Moreover, by (6.9)(ii), $[W_\tau, W_{T_\eta -}] \subseteq R \cup (K_+ \setminus E)$. As above, the case $W_{T_\eta} < W_{T_\eta -}$ violates the right-hand inequality of (6.28) or (6.26). Now let $W_{T_\eta} > W_{T_\eta -}$. First we assume that there is a right neighbourhood of $W_{T_\eta -}$ included in S_+. Then $\max_{s \leq T_\eta -} W_s > W_{T_\eta -}$ contradicts to the left-hand inequality of (6.28) whereas $\max_{s \leq T_\eta -} W_s = W_{T_\eta -}$ leads to a contradiction of (6.26). If there is not a right neighbourhood of $W_{T_\eta -}$ included in S_+ then, using (2.11)[2], we can find an interval $(a_n, b_n) \subseteq (W_{T_\eta -}, W_{T_\eta})$ with $a_n \in K_+ \setminus E$. As above, in the case $W_{T_\eta -} = a_n \in K_+ \setminus E$, $W_{T_\eta} > a_n$, this leads again to a contradiction. This ends the proof of $W_{T_\eta} = W_{T_\eta -}$ for $W_{T_\eta -} \in S_+$.

Finally, the cases $W_{T_\eta -} = b_n \in K_- \setminus E$ and $W_{T_\eta -} \in S_-$ can be treated analogously, finishing the proof. □

Let us take a review of the status of the proof of (6.5). The properties (6.18), (6.16), (6.15), (6.22), (6.25) just cover the conditions on A^+ we need to apply (3.6) with respect to $(W, \mathbb{F}^W)$. Therefore, applying (3.6), we have shown that $(W \circ T, \mathbb{F}^W \circ T)$ is a continuous strong Markov process on $(\Omega, \mathcal{F}, \mathbf{P}_x, x \in$

[2] See also the proof of (2.11).

$\mathbb{R}$) with shift operators $\Theta \circ T$. In the following lemmata we verify the asserted properties of $X = W \circ T$.

(6.31) Lemma *Let $x \notin R^*$ be fixed. Then the identity*

$$\mathrm{D}_y(X) = \begin{cases} A^+_{\mathrm{D}_y(W)-} \mathbf{1}_{\{\mathrm{D}_y(W) < \mathrm{D}_{b_n}(W)\}} + \infty \cdot \mathbf{1}_{\{\mathrm{D}_y(W) \geq \mathrm{D}_{b_n}(W)\}} : & \begin{array}{l} y \in I_n \\ n \in N_{++} \end{array} \\ A^+_{\mathrm{D}_y(W)-} \mathbf{1}_{\{\mathrm{D}_y(W) < \mathrm{D}_{a_n}(W)\}} + \infty \cdot \mathbf{1}_{\{\mathrm{D}_y(W) \geq \mathrm{D}_{a_n}(W)\}} : & \begin{array}{l} y \in I_n \\ n \in N_{--} \end{array} \\ A^+_{\mathrm{D}_y(W)-} & : \textit{otherwise} \end{cases}$$

holds true $\mathbf{P}_x$-a.s. for all $y \in I(x)$.

Proof of (6.31). Simplifying the notation we set $\eta = \mathrm{D}_y(W)$.

To begin with we consider samples ω satisfying $\omega \in \{\eta < \mathrm{D}_{b_n}(W)\}$ if $y \in I_n, n \in N_{++}$, or $\omega \in \{\eta < \mathrm{D}_{a_n}(W)\}$ if $y \in I_n, n \in N_{--}$. Otherwise, $\omega \in \Omega$ can be arbitrary.

First let $A^+_\eta < \infty$ so that $\eta < T_\infty$ as well as $A^+_\eta = A^+_{\eta-}$ (see (6.16)). Combining $\mathbf{P}_y(\{\mathrm{D}_{y-}(W) = 0\}) = \mathbf{P}_y(\{\mathrm{D}_{y+}(W) = 0\}) = 1$, (6.4)(i) and (A1.10), we obtain $A^+_{\eta+\varepsilon} > A^+_\eta$ $\mathbf{P}_x$-a.s., $\varepsilon > 0$, showing $T_{A^+_\eta} = \eta$. All in all,

$$\mathrm{D}_y(X) = A^+_\eta = A^+_{\eta-} \quad \text{on} \quad \{A^+_\eta < \infty\} \qquad \mathbf{P}_x - \text{a.s.}$$

holds.

Next, assume $A^+_\eta = \infty$, $A^+_{\eta-} < \infty$ implying $T_t = \eta$ for all $t \geq A^+_{\eta-}$ as well as $T_t < \eta$ if $t < A^+_{\eta-}$. But this yields $\mathrm{D}_y(X) = A^+_{\eta-}$ on $\{A^+_\eta = \infty, A^+_{\eta-} < \infty\}$.

Finally, let $A^+_\eta = A^+_{\eta-} = \infty$. Then we have $T_t < \eta$ for all t proving $\mathrm{D}_y(X) = \infty = A^+_{\eta-}$ on $\{A^+_\eta = A^+_{\eta-} = \infty\}$.

So, it remains to show that $\mathrm{D}_y(X) = \infty$ $\mathbf{P}_x$-a.s. on $\{\eta \geq \mathrm{D}_{b_n}(W)\}$ resp. on $\{\eta \geq \mathrm{D}_{a_n}(W)\}$ in the cases $y \in I_n, n \in N_{++}$, and $y \in I_n, n \in N_{--}$, respectively. By symmetry, it suffices to treat the first case.

For $\omega \in \{\eta \geq \mathrm{D}_{b_n}(W)\}$ from (6.7) it follows that $A^+_{\eta(\omega)}(\omega) = A^+_{\eta(\omega)+\varepsilon(\omega)}(\omega)$, $\varepsilon(\omega) > 0$ sufficiently small. Moreover, $\{(t,\omega) : W_t(\omega) = y, \eta(\omega) \geq D_{b_n}(W)(\omega)\}$ $\cap H_+(A^+) = \emptyset$. Consequently, in view of $[\![T_t]\!] \subseteq H_+(A^+) \cup [\![T_\infty]\!]$, $, t \geq 0$, we obtain $\{t : (W \circ T)_t = y\} = \emptyset$ on $\{\eta \geq \mathrm{D}_{b_n}(W)\}$, i.e. $\mathrm{D}_y(X) = \infty$ on this set. □

(6.32) Lemma *The set of regular, right and left singular points of $X = W \circ T$ is the set R, K_+ and K_-, respectively.*

Proof of (6.32). First let $x \in R$, $x \in (a_n, b_n)$. Combining (6.4)(ii) and (A1.9)(ii),

$$A_t < \infty, \qquad t < \mathrm{D}_{a_n}(W) \wedge \mathrm{D}_{b_n}(W), \mathbf{P}_x - \text{a.s.}$$

holds. Moreover, using (6.4)(i) and (A1.10), A is strictly increasing as well as continuous on $[0, \mathrm{D}_{a_n}(W) \wedge \mathrm{D}_{b_n}(W))$ a.s. Clearly, this yields $\mathrm{D}_{x+}(X) = A_{\mathrm{D}_{x+}(W)}$ $\mathbf{P}_x$-a.s. implying

$$\mathbf{P}_x(\{\mathrm{D}_{x+}(X) = 0\}) = 1$$

since W is a regular process. Analogously we obtain $\mathbf{P}_x(\{\mathrm{D}_{x-}(X) = 0\}) = 1$. This means that x is regular for X.

Now let $x \in E = K_+ \cap K_-$. Then from (6.17) it follows that $T_t = 0$, $t \geq 0$, $\mathbf{P}_x$-a.s., implying that $x \in E$ is absorbing for X.

Finally, let $x \in K_+ \setminus E$. By symmetry, the case $x \in K_- \setminus E$ can be treated analogously. Here we have $\tau = 0$ as well as $f_+(W_\tau) = f_+(x) = x$ $\mathbf{P}_x$-a.s. But $\tau = 0$ and $[\![T_t]\!] \subseteq H_+(A^+) \cup [\![T_\infty]\!]$ give $X = W \circ T \subseteq I(W_\tau) = I(x)$ $\mathbf{P}_x$-a.s., i.e. $\mathbf{P}_x(\{\mathrm{D}_{x-}(X) = 0\}) = 1$. Hence, x is right singular for X. In order to verify that x is not absorbing for X we show that $\mathbf{P}_x(\{\mathrm{D}_y(X) < \infty\}) > 0$ for some $y > x$. Since K_- is assumed to be closed to the left-hand side, from $x \in K_+ \setminus E$ follows that $f_-(x) > x$. (6.17) implies that there exists a point y satisfying $x < y < f_-(x)$ and $\mathbf{P}_x(\{A^+_{\mathrm{D}_y(W)} < \infty\}) > 0$. By (6.31) this yields $\mathbf{P}_x(\{\mathrm{D}_y(X) < \infty\}) > 0$. □

(6.33) Lemma *The function $p(x) = x$, $x \in \mathbb{R}$, is a scale function of the process X.*

Proof of (6.33). We fix $x \in (a_n, b_n) \subseteq R$ and choose arbitrary points y, z satisfying $a_n < y < x < z < b_n$. Recalling the proof of (6.32), A^+ is strictly increasing and continuous on $[0, \mathrm{D}_{a_n}(W) \wedge \mathrm{D}_{b_n}(W))$ $\mathbf{P}_x$-a.s. This implies $\mathrm{D}_y(X) = A^+_{\mathrm{D}_y(W)}$, $\mathrm{D}_z(X) = A^+_{\mathrm{D}_z(W)}$ $\mathbf{P}_x$-a.s. Therefore,

$$\begin{aligned}
\mathbf{P}_x(\{\mathrm{D}_y(X) < \mathrm{D}_z(X)\}) &= \mathbf{P}_x(\{A^+_{\mathrm{D}_y(W)} < A^+_{\mathrm{D}_z(W)}\}) \\
&= \mathbf{P}_x(\{\mathrm{D}_y(W) < \mathrm{D}_z(W)\}) \\
&= (z - x)/(z - y),
\end{aligned}$$

because $p(x) = x$, $x \in \mathbb{R}$, is a scale function of W. □

We accomplish the proof of (6.5) by showing that $m|_{I \setminus E}$ just represents the speed measure of X associated with p. For that purpose we prove a technical lemma. Observe that by (1.21), $X \in \mathcal{S}(\mathbb{F}^X)$. Hence the local times of X are well-defined.

(6.34) Lemma *Let $x \notin R^*$.*

(i) *For $n \in N_{+-}$, if $I_n \subseteq I(x)$ then $\mathbf{P}_x$-a.s.*

$$L^W(T_t, y) = L^X(t, y), \qquad t \geq 0,\ y \in (a_n, b_n),$$

$$\frac{1}{2} L^W(T_t, y) = L^X(t, y), \qquad t \geq 0,\ y = a_n, b_n.$$

(ii) *For $n \in N_{++}$, if $I_n \subseteq I(x)$ then $\mathbf{P}_x$-a.s.*

$$L^W(T_t \wedge \mathrm{D}_{b_n}(W), y) = L^X(t, y), \qquad t \geq 0,\ y \in (a_n, b_n),$$

$$\frac{1}{2} L^W(T_t \wedge \mathrm{D}_{b_n}(W), a_n) = L^X(t, a_n), \qquad t \geq 0.$$

A corresponding assertion is true for $n \in N_{--}$.

Proof of (6.34). First we remark that, by (6.32),(1.17)(iv), for $a_n \in K_+$ (resp. $b_n \in K_-$) we have

(6.35) $$L^X_-(\cdot, a_n) = 0 \quad (\text{resp. } L^X_+(\cdot, b_n) = 0) \quad \text{a.s.}$$

In what follows we apply (1.17)(v) using the notation introduced there.

(i) Let $n \in N_{+-}$ with $I_n \subseteq I(x)$ and let $t \geq 0$, $y \in [a_n, b_n)$ be fixed. Then by (1.17)(v) for all $N \in \mathbb{N}$ we have the convergence

$$\varepsilon C^\varepsilon_{T_t}(W) \cdot \mathbf{1}_{\{T_t \leq N\}} \to \frac{1}{2} L^W(T_t, y) \cdot \mathbf{1}_{\{T_t \leq N\}}$$

in $\mathbf{P}_x$-probability as ε tends to zero. For the process X (1.17)(v) also guarantees the convergence

$$\varepsilon C^\varepsilon_t(X) \to \frac{1}{2} L^X_+(t, y),\ \varepsilon \to 0,$$

in $\mathbf{P}_x$-probability. Now, for each $\varepsilon > 0$ we will show that $\mathbf{P}_x$-a.s.

(6.36) $$C^\varepsilon_{T_t}(W) = C^\varepsilon_t(X).$$

As a consequence of (6.36), taking into account that $T_t < \infty$ $\mathbf{P}_x$-a.s. (cf. (6.15)) we obtain

$$L^X_+(t, y) = L^W(T_t, y) \qquad \mathbf{P}_x - \text{a.s.}$$

Here the $\mathbf{P}_x$-exceptional set does not depend on (t, y) because both sides are right-continuous in (t, y).

So (6.36) remains to be proved. For the stopping times introduced in (1.17)(v), as in the proof of (6.31), we establish

$$V^\varepsilon_k(X) = A^+_{V^\varepsilon_k(W)-},\ U^\varepsilon_k(X) = A^+_{U^\varepsilon_k(W)-} \qquad \mathbf{P}_x - \text{a.s.}$$

Using $\{s \leq T_t\} = \{A^+_{s-} \leq t\}$ (see (6.16)), we get

$$C^\varepsilon_{T_t}(W) = \sum_{k \geq 1} \mathbf{1}_{\{U^\varepsilon_k(W) \leq T_t\}} = \sum_{k \geq 1} \mathbf{1}_{\{A^+_{U^\varepsilon_k(W)-} \leq t\}} = C^\varepsilon_t(X)$$

for all $t \geq 0$ $\mathbf{P}_x$-a.s.

Analogously, applying (6.35) and (1.13), one obtains $\mathbf{P}_x$-a.s.

$$L^X_-(t, y) = L^W(T_t, y), \qquad t \geq 0,\ y \in (a_n, b_n],$$

and the assertion of (i) follows.

(ii) Fix $t \geq 0$ and $y \in [a_n, b_n)$, $n \in N_{++}$ with $I_n \subseteq I(x)$. Clearly, if we can show that

$$C^{\varepsilon}_{T_t \wedge \mathrm{D}_{b_n}(W)}(W) = C^{\varepsilon}_t(X) \qquad \mathbf{P}_x - \text{a.s.}$$

for all sufficiently small $\varepsilon > 0$ then we can repeat the proof of (i) proving the assertion.

Indeed, the desired equality is an easy consequence of

$$V^{\varepsilon}_k(X) = \begin{cases} A^+_{V^{\varepsilon}_k(W)-} & : \quad V^{\varepsilon}_k(W) \leq \mathrm{D}_{b_n}(W) \\ \infty & : \quad \text{otherwise} \end{cases}$$

$$U^{\varepsilon}_k(X) = \begin{cases} A^+_{U^{\varepsilon}_k(W)-} & : \quad U^{\varepsilon}_k(W) \leq \mathrm{D}_{b_n}(W) \\ \infty & : \quad \text{otherwise} \end{cases}$$

which can be verified as in (6.31) if $y + \varepsilon < b_n$. □

(6.37) Lemma $m|_{I \setminus E}$ *is the speed measure of* X *with respect to* p.

Proof of (6.37). According to (5.30) we only have to show that

$$\textbf{(6.38)} \qquad t = \int_{\bar{R} \setminus E} L^X(t, y)\, m(dy) + m([X_0, \max_{s \leq t} X_s] \cap S_+) + m([\min_{s \leq t} X_s, X_0] \cap S_-)$$

for all $t < \mathrm{D}_E$ a.s. To verify (6.38) we compute the right-hand side of (6.7) at the time T_t instead of t.

Combining (A1.9), (A1.10) and (6.4)(i),(ii) the process A is strictly increasing and continuous on $[0, \tau]$ so that

$$T_t < \tau \quad \text{if and only if} \quad t < A_\tau = \mathrm{D}_{(R^*)^c}(X).$$

We write $\tau(X) = \mathrm{D}_{(R^*)^c}(X)$ in what follows. Then (6.7) yields

$$\begin{aligned} t = A_{T_t} &= \int_{R^*} L^W(T_t, y)\, m(dy) \\ &= \int_{R^*} L^X(t, y)\, m(dy) \\ \textbf{(6.39)} \qquad &= \int_{\bar{R} \setminus E} L^X(t, y)\, m(dy) + m([X_0, \max_{s \leq t} X_s] \cap S_+) \\ &\quad + m([\min_{s \leq t} X_s, X_0] \cap S_-) \end{aligned}$$

a.s. for all $t < \tau(X)$, where we have used that $L^W(T_t, y) = L^X(t, y)$ which follows from [32], (10.17), because T is continuous on $[0, A_\tau)$. On $\{\tau(X) > 0\}$ we have $\tau(X) < \mathrm{D}_E(X)$ and, by (6.4)(iii), the right-hand side of (6.39) is left-continuous on $[0,\tau(X))$. Consequently, (6.38) is shown for all $t \le \tau(X)$ a.s.

So (6.38) remains to be proved for all $\tau(X)+t$ on $\{\tau(X)<\infty\}\cap\{\tau(X)+t<\mathrm{D}_E(X)\}$, $t\ge 0$, a.s. Applying (1.16), (6.39), on $\{\tau(X)<\infty\}\cap\{\tau(X)+t<\mathrm{D}_E(X)\}$ we obtain

$$\begin{aligned}\Big(\int_{\bar R\setminus E} L^X(t,y)\,m(\mathrm{d}y) \;+\; m([X_0,\max_{s\le t}X_s]\cap S_+)\\ + m([\min_{s\le t}X_s,X_0]\cap S_-)\Big)\circ\theta_{\tau(X)} \;+\; \tau(X)\\ = \int_{\bar R\setminus E} L^X(t+\tau(X),y)\,m(\mathrm{d}y) \;+\; m([X_0,\max_{s\le t+\tau(X)}X_s]\cap S_+)\\ + m([\min_{s\le t+\tau(X)}X_s,X_0]\cap S_-),\end{aligned}$$

where we have also used (see (6.4)(iii))

$$m([X_0,\max_{s\le\tau(X)}X_s]\cap S_+) \;=\; m([\min_{s\le\tau(X)}X_s,X_0]\cap S_-) \;=\; 0.$$

Hence, using the strong Markov property of X, we have to verify (6.38) only for measures $\mathbf{P}_x$ with $x\notin R^*$.

Let $t\ge 0$ and $x\notin R^*$ be fixed. Then from (6.16) it follows that $t=A_{T_t}$ on $\{T_t<T_\infty\}$ $\mathbf{P}_x$-a.s. Now (6.7) gives

$$\begin{aligned} t \;=\;& \sum_{n\in N_{+-}} \mathbf{1}_{I_n\subseteq I(x)}\int_{I_n} L^W(T_t,y)\,\tilde m(\mathrm{d}y)\\ +& \sum_{n\in N_{++}} \mathbf{1}_{I_n\subseteq I(x)}\int_{I_n} L^W(T_t\wedge \mathrm{D}_{b_n}(W),y)\,\tilde m(\mathrm{d}y)\\ +& \sum_{n\in N_{--}} \mathbf{1}_{I_n\subseteq I(x)}\int_{I_n} L^W(T_t\wedge \mathrm{D}_{a_n}(W),y)\,\tilde m(\mathrm{d}y)\\ +& \;\tilde m([\min_{s\le T_t}W_s,\max_{s\le T_t}W_s]\cap I(x)\cap \bar R^c)\end{aligned}$$

on $\{T_t<T_\infty\}$ $\mathbf{P}_x$-a.s. Using (6.34) and (6.6), this leads to

$$\begin{aligned} t \;=\;& \sum_{n\in N_{+-}\cup N_{++}\cup N_{--}} \mathbf{1}_{I_n\subseteq I(x)}\int_{I_n} L^X(t,y)\,m(\mathrm{d}y)\\ +&\; m([\min_{s\le T_t}W_s,\max_{s\le T_t}W_s]\cap I(x)\cap \bar R^c)\end{aligned}$$

on $\{T_t<T_\infty\}$ $\mathbf{P}_x$-a.s. In view of $[\![T_t]\!]\subseteq H_+(A^+)\cup[\![T_\infty]\!]\subseteq\{W\in I(x)\}$ $\mathbf{P}_x$-a.s. we have $X_t\in I(x)$ $\mathbf{P}_x$-a.s. and, as a consequence, combining $I(x)\cap R^*=\emptyset$ and (1.17)(i),

$$L^X(t,y)=0, \qquad\qquad y\in R^*,\ \mathbf{P}_x-\text{a.s.}$$

Further, $I_n \cap I(x) \subseteq E$ holds for intervals I_n with $I_n \cap I(x) \neq \emptyset$, $I_n \not\subseteq I(x)$. But (1.17)(ii) and (2.3)(iii) imply

$$L^X(t,y) = 0, \qquad y \in E, \mathbf{P}_x - \text{a.s.}$$

All in all, we obtain

$$\text{(6.40)} \quad t = \int_{\bar{R}\setminus E} L^X(t,y)\, m(dy) + m([\min_{s\le T_t} W_s, \max_{s\le T_t} W_s] \cap I(x) \cap \bar{R}^c)$$

on $\{T_t < T_\infty\}$ $\mathbf{P}_x$-a.s.

Now we are going to show that

$$\text{(6.41)} \quad m([x, \max_{s\le T_t} W_s] \cap I(x) \cap \bar{R}^c) = m([x, \max_{s\le t} X_s] \cap I(x) \cap \bar{R}^c)$$

on $\{T_t < T_\infty\}$ $\mathbf{P}_x$-a.s. First of all, obviously

$$[x, \max_{s\le t} X_s] \subseteq [x, \max_{s\le T_t} W_s].$$

Moreover

$$\text{(6.42)} \quad [x, \max_{s\le T_t} W_s] \setminus \bigcup_{u\le t} (\max_{s\le T_{u-}} W_s, \max_{s\le T_u} W_s] \subseteq [x, \max_{s\le t} X_s].$$

In fact, if $y \in [x, \max_{s\le T_t} W_s]$ satisfies $y \notin [x, \max_{s\le t} X_s]$ then there exists some $u \le t$ such that $D_y(W) \in (T_{u-}, T_u)$ and, consequently,

$$\max_{s\le T_{u-}} W_s < y \le \max_{s\le T_u} W_s$$

that is (6.42).

(6.9)(ii) implies $T_t < \inf\{s \ge 0 : W_s \in E \cap I(x) \cap \bar{R}^c\}$ on $\{T_t < T_\infty\}$ $\mathbf{P}_x$-a.s. Therefore, (6.4)(iii) yields $m(\{W_s\} \cap I(x) \cap \bar{R}^c) = 0$ for all $s \le T_t$ on $\{T_t < T_\infty\}$ $\mathbf{P}_x$-a.s. Since A is constant on each $[T_{u-}, T_u]$,

$$m(\bigcup_{u\le t} (\max_{s\le T_{u-}} W_s, \max_{s\le T_u} W_s] \cap I(x) \cap \bar{R}^c) = 0$$

on $\{T_t < T_\infty\}$ $\mathbf{P}_x$-a.s.[3] In view of (6.42) this gives (6.41) and, analogously,

$$\text{(6.43)} \quad m([\min_{s\le T_t} W_s, x] \cap I(x) \cap \bar{R}^c) = m([\min_{s\le t} X_s, x] \cap I(x) \cap \bar{R}^c)$$

on $\{T_t < T_\infty\}$ $\mathbf{P}_x$-a.s.

Combining (6.40), (6.41), (6.43), $X_t \in I(x)$ $\mathbf{P}_x$-a.s. and the definition of $I(x)$ we obtain

$$t = \int_{\bar{R}\setminus E} L^X(t,y)\, m(dy) + m([x, \max_{s\le t} X_s] \cap \bar{R}^c) + m([\min_{s\le t} X_s, x] \cap \bar{R}^c)$$

[3] At most a countable number of sets in the union are nonempty.

on $\{T_t < T_\infty\}$ $\mathbf{P}_x$-a.s. But on $\{T_t < T_\infty\}$ we have $\mathbf{P}_x$-a.s. $t < \mathrm{D}_E(X)$ by (6.9)(ii). Vice versa, the strong Markov property of X yields $t \geq \mathrm{D}_E(X)$ on $\{T_t = T_\infty\}$ $\mathbf{P}_x$-a.s. Thus, we also have shown (6.38) for all $t < \mathrm{D}_E(X)$ $\mathbf{P}_x$-a.s., $x \notin R^*$, completing the proof. □

Now, all properties of $X = W \circ T$ asserted in (6.5) have been shown and the proof of (6.5) is complete. □

Now we depart from the assumptions $I = \mathbb{R}$, $p(x) = x$, $x \in I$, made in (6.5) and investigate the general case.

We start with an arbitrary interval $I \subseteq \mathbb{R}$ and subsets R, K_+, K_- as in (6.1). Further, let $p : I \to \mathbb{R}$ be a continuous strictly increasing function and let m be a p-admissible measure (see (6.4)). Denote by $q : p(I) \to I$ the inverse function to p.

We set

$$I' = \mathbb{R},\ p'(x) = x,\ x \in I',\ R' = p(R),$$

$$K'_+ = p(K_+) \cup (\mathbb{R} \setminus p(I)),\ K'_- = p(K_-) \cup (\mathbb{R} \setminus p(I)),$$

$$m'(A) = m \circ q(A \cap p(I)) + \infty \cdot \mathbf{1}_{A \setminus p(I) \neq \emptyset},\ A \in \mathfrak{B}(\mathbb{R}),$$

where $\mathbf{1}_{A \setminus p(I) \neq \emptyset}$ is equal to 1 if $A \setminus p(I) \neq \emptyset$ and 0 otherwise. It is easy to verify that m' is a p'-admissible measure on $(\mathbb{R}, \mathfrak{B}(\mathbb{R}))$.

(6.44) Proposition *Let $(X', \mathbb{F})$ on $(\Omega, \mathcal{F}, \mathbf{P}_x, x \in \mathbb{R})$ denote the continuous strong Markov process constructed as in (6.5) admitting as the set of regular, left singular, right singular points R', K'_-, K'_+, respectively, and with the speed measure $m'|_{\mathbb{R}\setminus E}$ associated with the scale function p'. Then, $X' \in p(I)$ holds $\mathbf{P}_x$-a.s. for all $x \in p(I)$. Moreover, $X = q(X')$ is a continuous strong Markov process on $(\Omega, \mathcal{F}, \mathbf{P}_{p(x)}, x \in I)$ taking values in I possessing R, K_-, K_+ as the set of regular, left singular, right singular points, respectively, and with $m|_{I\setminus E}$ as the speed measure associated with the scale function p.*

Proof. We fix $x' \in p(I)$ and show that

(6.45) $$\mathbf{P}_{x'}(\{X' \in p(I)\}) = 1.$$

Denote $z' = \sup p(I)$. If $z' = \infty$ then obviously $\mathbf{P}_{x'}(\{X' < z'\}) = 1$. If $\infty > z' \in p(I)$ then $z = q(z') = \sup I \in I$ and, from the assumptions on I, K_+, K_- made in (6.1), it follows that $z \in K_-$. Hence, z' is an element of K'_- implying

(6.46) $$\mathbf{P}_{x'}(\{X' \leq z'\}) = 1.$$

For $\infty > z' \notin p(I)$ we will show that

(6.47) $$\mathbf{P}_{x'}(\{X' < z'\}) = 1.$$

Obviously, this is true if $[x', z') \cap K'_- \neq \emptyset$. In the case of $[x', z') \cap K'_- = \emptyset$ we distinguish the cases $x' \in K'_+ \setminus E'$ and $x' \in R'$.

First let $x' \in K'_+ \setminus E'$. Then (2.9) yields $\mathbf{P}_{x'}(\{\mathrm{D}_{z'}(X') < \infty\}) \in \{0, 1\}$ and in the case of $\mathbf{P}_{x'}(\{\mathrm{D}_{z'}(X') < \infty\}) = 1$ it even holds that $\mathbf{E}_{x'}\mathrm{D}_{z'}(X') < \infty$. Let us assume that $\mathbf{P}_{x'}(\{\mathrm{D}_{z'}(X') < \infty\}) = 1$. We will show that this leads to a contradiction, proving (6.47) for the case of $x' \in K'_+ \setminus E'$.

For $\mathbf{E}_{x'}\mathrm{D}_{z'}(X') < \infty$, (5.27) and (5.36) imply

$$\begin{aligned}
\infty \;>\; \mathbf{E}_{x'}\mathrm{D}_{z'}(X') &= \mathbf{E}_{x'} \int_0^{\mathrm{D}_{z'}} \mathbf{1}_{[x', z')}(X'_s)\, \mathrm{d}s \\
&= \mathbf{E}_{x'} \int_{\bar{R}' \cap [x', z')} L^{X'}(\mathrm{D}_{z'}, y)\, m'(\mathrm{d}y) + m'([x', z') \cap S'_+) \\
&\geq \int_{[x', z')} \sum \mathbf{1}_{[a'_n, b'_n)}(y)(b'_n \wedge z' - y)\, m'(\mathrm{d}y) + m'([x', z') \cap S'_+) \\
&= \int_{[x, z)} \sum \mathbf{1}_{[a_n, b_n)}(y)(p(b_n) - p(y))\, m(\mathrm{d}y) + m([x, z) \cap S_+)
\end{aligned}$$

where $a'_n = p(a_n)$, $b'_n = p(b_n)$, $x = q(x')$ and $z = \sup I$. This inequality contradicts (6.4)(v).

Now let $x' \in R'$ that is $x' \in (a'_n, b'_n)$ for some n. If $b'_n < z'$ then from our assumption follows $b'_n \in K'_+ \setminus E'$. From the previous part of the proof, replacing x' by b'_n, we get (6.47) again. Now, let $b'_n = z'$. If $a'_n \in K'_-$, then $\mathbf{P}_{x'}(\{\mathrm{D}_{z'}(X') < \infty, \mathrm{D}_{a'_n}(X') < \mathrm{D}_{b'_n}(X')\}) = 0$ holds. On $\{\mathrm{D}_{a'_n}(X') > \mathrm{D}_{b'_n}(X')\}$, $a'_n \in K'_-$, or in the case $a'_n \in K'_+ \setminus E'$ (6.31) gives $\mathbf{P}_{x'}$-a.s.

$$\begin{aligned}
\mathrm{D}_{z'}(X') &= A^+_{\mathrm{D}_{z'}(W)-} \\
&\geq \int_{[x', b'_n)} L^W(\mathrm{D}_{b'_n}(W), y)\, m'(\mathrm{d}y)
\end{aligned}$$

where A denotes the process used for the construction of X' in (6.5). Now, by (A1.8), the right-hand side of the above inequality diverges $\mathbf{P}_{x'}$-a.s. if

$$\int_{[x', b'_n)} (b'_n - y)\, m'(\mathrm{d}y) = \infty.$$

But the latter is satisfied by (6.4)(v) and

$$\int_{[x', b'_n)} (b'_n - y)\, m'(\mathrm{d}y) = \int_{[x, b_n)} (p(b_n) - p(y))\, m(\mathrm{d}y).$$

Thus we have verified (6.47).

Analogously, for $z' = \inf p(I)$, we obtain assertions corresponding with (6.46) and (6.47). This finishes the proof of (6.45).

From (6.45) we deduce that $X = q(X') \in I$ $\mathbf{P}_{p(x)}$-a.s. for all $x \in I$. Further, it is easy to check the strong Markov property of $(X, \mathbb{F})$ on

$(\Omega, \mathcal{F}, \mathbf{P}_{p(x)}, x \in I)$ as well as the asserted properties of R, K_+, K_-. After all, combining (5.27) applied for X', (5.30) and $p(X) = X'$, $m|_{I \setminus E}$ is the speed measure of X associated with the scale function p. □

The following theorem is an immediate consequence of (6.44).

(6.48) Theorem *For an arbitrary interval $I \subseteq \mathbb{R}$ with subsets R, K_+, K_- as in (6.1), an arbitrary function $p : I \to \mathbb{R}$ according to (6.3) as well as for each p-admissible measure m there exists a continuous strong Markov process taking values in I admitting R, K_+, K_- as the set of regular, right singular, left singular points, respectively, and $m|_{I \setminus E}$ as the speed measure associated with the scale function p.*

As already pointed out, by (5.34), the conditions (6.4)(i)-(v) are necessary for $m|_{I \setminus E}$ being the speed measure of a continuous strong Markov process with given scale function p. On the other hand, every measure n on $(I \setminus E, \mathfrak{B}(I \setminus E))$ satisfying (6.4)(i)-(v) can be extended to a p-admissible measure on $(I, \mathfrak{B}(I))$, for example, by

$$m(A) = n(A \cap I \setminus E) + \infty \cdot \mathbf{1}_{A \cap E \neq \emptyset}, \qquad A \in \mathfrak{B}(I).$$

(6.49) Remark However, in general there is not only one p-admissible measure m satisfying $m|_{I \setminus E} = n$. This is also a reason why we have introduced the speed measure as a measure on $(I \setminus E, \mathfrak{B}(I \setminus E))$. One could also define a speed measure which lives on the "whole" measure space $(I, \mathfrak{B}(I))$ by identifying all suitable measures being identical on $\mathfrak{B}(I \setminus E)$ and satisfying (6.4)(vi). This approach was taken in [21] for continuous strong Markov martingales.

As a consequence, the above comments lead to the following

(6.50) Proposition *Let us be given an interval $I \subseteq \mathbb{R}$ decomposed into subsets R, K_+, K_- as in (6.1) and let p be a function $p : I \to \mathbb{R}$ according to (6.3). Then for every measure m on $(I \setminus E, \mathfrak{B}(I \setminus E))$ satisfying (6.4)(i)-(v) there exists a continuous strong Markov process taking values in I with R, K_+, K_- as the set of regular, right singular, left singular points, respectively, and possessing m as the speed measure associated with p.*

VII. Continuous Strong Markov Semimartingales as Solutions of Stochastic Differential Equations

(7.1) Let $(X, \mathbb{F})$ on $(\Omega, \mathcal{F}, \mathbf{P}_x, x \in I)$ be a continuous strong Markov semimartingale with speed measure m associated with a given scale function p (cf. (5.13)) and sets R, K_+, K_- as the set of regular, right singular, left singular points, respectively. As in (1.10), for the semimartingale decomposition of X we use the notation

$$X = X_0 + M(X) + V(X)$$

with $M(X) \in \mathcal{M}(\mathbb{F}^X)$, $V(X) \in \mathcal{V}(\mathbb{F}^X)$.

Further, we denote by d_+ (resp. d_-) the S_+-drift function (resp. S_--drift function) of X (cf. (4.27)). Define $g : \mathbb{R} \to [0, \infty)$ by (4.4) and recall that g has locally bounded variation on each component of R (cf. (4.2)(ii)). Finally, we set $\bar{g}(x) = \frac{1}{2}(g(x) + g(x-))$ for $x \in R$ and extend $\bar{g}$ to $\bar{R}$ as in (4.6).

In this chapter we investigate the problem, under which conditions $(X, \mathbb{F})$ turns out to be a solution of a certain stochastic differential equation. Here we only deal with stochastic differential equations driven by a Wiener process and, as a consequence, the Lebesgue decomposition of the speed measure m will play an outstanding role.

By (5.34)(iv), m is σ-finite; therefore it possesses a Lebesgue decomposition, i.e. there exist a set $N \in \mathfrak{B}(I \setminus E)$, a measure n on $\mathfrak{B}(I \setminus E)$ and a $\mathfrak{B}(I \setminus E)$-measurable function $h : I \setminus E \to [0, \infty]$ such that for all $A \in \mathfrak{B}(I \setminus E)$

$$m(A) = \int_A h(x)\,\mathrm{d}x + n(A) \tag{7.2}$$

where $n((I \setminus E) \setminus N) = \ell(N) = 0$. Without restricting the generality we set

$$h(x) = \infty \quad \text{for} \quad x \in N. \tag{7.3}$$

VII.1 Equations with Generalized Drift

(7.4) Let $I \subseteq \mathbb{R}$ denote a non-empty interval. Suppose we are given the following objects:

- a $\mathfrak{B}^u(I)$-measurable function $b : I \to \bar{\mathbb{R}}$,
- an $\mathbb{R}$-open subset $R \subseteq I$, $R = \bigcup_n (a_n, b_n)$,

- a right-continuous function $g : R \to [0, \infty)$ which is of locally bounded variation on each component (a_n, b_n) of R,
- at most countable sets $B_+, B_- \subseteq I$,
- measurable sets $A_+, A_- \subseteq I$,
- $\mathfrak{B}^u(I) \otimes \mathfrak{B}(\mathbb{R}_+)$-measurable functions $a_+, a_- : I \times \mathbb{R}_+ \to [0, \infty)$ being continuous and increasing in the second variable.

We investigate the following stochastic differential equation

$$\begin{aligned} X_t \;=\; & X_0 + \int_0^t b(X_s)\,\mathrm{d}B_s + \frac{1}{2}\int_R L^X(t,y)\,\mu_g(\mathrm{d}y) \\ & + \sum_{x\in B_+} L^X(t,x) - \sum_{x\in B_-} L^X(t,x) \qquad\qquad (7.5) \\ & + a_+\left(X_0, \int_0^t \mathbf{1}_{A_+}(X_s)\,\mathrm{d}s\right) - a_-\left(X_0, \int_0^t \mathbf{1}_{A_-}(X_s)\,\mathrm{d}s\right) \end{aligned}$$

where L^X is the symmetric local time of the semimartingale X and $(B_t)_{t\geq 0}$ denotes a Wiener process. For the definition of the integral with respect to μ_g we refer to (A2.7).

(7.6) Definition An I-valued continuous random process X defined on a family of probability spaces $(\Omega, \mathcal{F}, \mathbf{P}_x, x \in I)$ as in (1.1) equipped with a filtration $\mathbb{F}$ according to (1.2) is said to be a *solution of equation* (7.5) if

(i) $X \in \mathcal{S}(\mathbb{F})$;
(ii) There exists a Wiener process $(B, \mathbb{F})$ on $(\Omega, \mathcal{F}, \mathbf{P}_x, x \in I)$ (cf. (1.11)) such that (7.5) holds for all $t \geq 0$ $\mathbf{P}_x$-a.s. for all $x \in I$.

We call equation (7.5) a stochastic differential equation with *generalized drift*.

(7.7) Remark (7.6)(ii) also requires that the stochastic integral in (7.5) is well defined for all $\mathbf{P}_x$, $x \in I$, that the integral with respect to μ_g exists (see (A2.7)) and that the sums in (7.5) are absolutely converging. Further, by (1.6) there exists a version of the local time of X which does not depend on the underlying probability measure $\mathbf{P}_x$.

Now let $(X, \mathbb{F})$ on $(\Omega, \mathcal{F}, \mathbf{P}_x, x \in I)$ be a continuous strong Markov semimartingale according to (7.1) and with Lebesgue decomposition of its speed measure given by (7.2),(7.3).

(7.8) Proposition *It holds that*

$$\mathrm{d}\langle M(X)\rangle_t \ll \mathrm{d}t \qquad a.s.$$

if and only if

$$h > 0 \quad \text{on} \quad \{x \in R : \bar{g}(x) > 0\} \quad \textit{Lebesgue-almost everywhere.}$$

In the case of $\langle M(X)\rangle$ being absolutely continuous we have

$$\langle M(X)\rangle_t = \int_0^t h^{-1}(X_s)\bar{g}(X_s)\mathbf{1}_R(X_s)\,\mathrm{d}s, \qquad t \geq 0, \text{ a.s.}$$

Proof. Using the occupation time formula (5.19) we first show that the condition on h is sufficient for the absolute continuity of $\langle M(X)\rangle$. Taking into account that $\ell(N)=0$, that is $h^{-1}=0$ on N[1], this formula yields

$$\begin{aligned}
&\int_0^t \mathbf{1}_R(X_s)h^{-1}(X_s)\bar{g}(X_s)\,\mathrm{d}s \\
&\quad = \int_{\{\bar{g}>0\}\setminus N} L^X(t,y)(\bar{g}(y))^{-1}\mathbf{1}_R(y)h^{-1}(y)\bar{g}(y)h(y)\,\mathrm{d}y \\
&\qquad + \int_0^t \mathbf{1}_R(X_s)h^{-1}(X_s)\bar{g}(X_s)\mathbf{1}_{\{\bar{g}=0\}}\,\mathrm{d}s \\
&\quad = \int_{\{\bar{g}>0\}\cap R} L^X(t,y)\,\mathrm{d}y \\
&\quad = \int_R L^X(t,y)\,\mathrm{d}y
\end{aligned}$$

for all $t \geq 0$ a.s. where the last equality is due to (4.7). Applying (1.15) and (4.29) we now obtain

$$\int_0^t \mathbf{1}_R(X_s)h^{-1}(X_s)\bar{g}(X_s)\,\mathrm{d}s = \int_0^t \mathbf{1}_R(X_s)\,\mathrm{d}\langle M(X)\rangle_s = \langle M(X)\rangle_t$$

for all $t \geq 0$ a.s.

Let us turn to the proof of the necessity of the stated condition and assume that $\mathrm{d}\langle M(X)\rangle_t \ll \mathrm{d}t$. Recall that $\langle M(X)\rangle$ is additive (see (1.8)). Hence, all assumptions of (2.23) are satisfied and there exists a $\mathfrak{B}^u(I)$-measurable function f such that

$$\langle M(X)\rangle_t = \int_0^t f(X_s)\,\mathrm{d}s, \qquad t \geq 0, \text{a.s.}$$

Let $k: I \to [0,\infty)$ be an arbitrary measurable function. Combining (1.15), (5.19) and $\ell(N)=0$, it follows that

$$\begin{aligned}
&\int_{\{\bar{g}>0\}} \mathbf{1}_R(y)L^X(t,y)k(y)\,\mathrm{d}y \\
&\quad = \int_0^t \mathbf{1}_{\{\bar{g}>0\}\cap R\setminus N}(X_s)k(X_s)\,\mathrm{d}\langle M(X)\rangle_s \\
&\quad = \int_0^t \mathbf{1}_{\{\bar{g}>0\}\cap R\setminus N}(X_s)k(X_s)f(X_s)\,\mathrm{d}s \\
&\quad = \int_{\{\bar{g}>0\}} \mathbf{1}_R(y)L^X(t,y)k(y)f(y)\bar{g}^{-1}(y)h(y)\,\mathrm{d}y.
\end{aligned}$$

[1] Remind (1.28).

Consequently, for $k = \mathbf{1}_{\{h=0\}}$ we get

$$\int_{\{h=0\}\cap\{\bar{g}>0\}\cap R} L^X(t,y)\,\mathrm{d}y = 0, \qquad t \geq 0, \text{ a.s.,}$$

leading to $\ell(\{h=0\}\cap\{\bar{g}>0\}\cap R) = 0$ by (4.7). □

Let us return to the semimartingale decomposition $X = X_0 + M(X) + V(X)$. From (4.3) we know the explicit structure of $V(X)$. Now X is a solution of a stochastic differential equation as in (7.5) if the martingale part $M(X)$ admits the representation

$$M(X)_t = \int_0^t b(X_s)\,\mathrm{d}B_s$$

with a certain function $b : I \to \bar{\mathbb{R}}$. Clearly, for this to be true, we need $\mathrm{d}\langle M(X)\rangle_t \ll \mathrm{d}t$. Now (4.3) and (7.8) yield the following theorem.

(7.9) Theorem *Let $(X,\mathbb{F})$ on $(\Omega,\mathcal{F},\mathbf{P}_x, x \in I)$ be a continuous strong Markov semimartingale with speed measure m associated with a scale function p whose Lebesgue decomposition is given by (7.2), (7.3). Then there exists a $\mathfrak{B}^u(I)$-measurable function $b : I \to \bar{\mathbb{R}}$ such that $(X,\mathbb{F})$ is a solution of the stochastic differential equation*

$$\begin{aligned} X_t &= X_0 + \int_0^t b(X_s)\,\mathrm{d}B_s + \frac{1}{2}\int_R L^X(t,y)\,\mu_g(\mathrm{d}y) \\ &\quad + \sum_{a_n\in K_+} L^X(t,a_n) - \sum_{b_n\in K_-} L^X(t,b_n) \\ &\quad + d_+\left(X_0, \int_0^t \mathbf{1}_{S_+}(X_s)\,\mathrm{d}s\right) - d_-\left(X_0, \int_0^t \mathbf{1}_{S_-}(X_s)\,\mathrm{d}s\right) \end{aligned}$$

(see (4.3)) if and only if $h > 0$ on $\{x \in R : \bar{g}(x) > 0\}$ Lebesgue-almost everywhere. Here we can set

$$b = \sqrt{\mathbf{1}_R h^{-1}\bar{g}}.$$

(7.10) Remark In general our stochastic basis $(\Omega,\mathcal{F},\mathbf{P}_x, x \in I),\mathbb{F}$, may be too "small" to support a Wiener process adapted to $\mathbb{F}$. In this case we pass over to a suitable extension of $(\Omega,\mathcal{F},\mathbf{P}_x, x \in I),\mathbb{F}$ (see (1.25)).

Proof of (7.9). The assertion immediately follows from (7.8), (1.26) and (4.3). □

(7.11) Remark Stochastic differential equations with drift of the form

$$\frac{1}{2}\int L^X(t,y)\,\nu(\mathrm{d}y)$$

where ν denotes a signed measure have already been investigated in [20], [23], [37]. In [20], [23] one finds necessary and sufficient conditions for the existence as well as for the uniqueness of those equations. The stochastic differential equations (7.5) considered here are more general because new terms are added and, on the other hand, the integrator μ_g in

$$\frac{1}{2}\int L^X(t,y)\,\mu_g(\mathrm{d}y)$$

need not to be a signed measure on the components of R (see Appendix 2). As far as we know, a systematic study of stochastic differential equations of the form (7.5) has not yet been accomplished.

VII.2 Equations with Ordinary Drift

Starting with an I-valued continuous strong Markov semimartingale $(X,\mathbb{F})$, in this section we give a necessary and sufficient condition such that $(X,\mathbb{F})$ turns out to be a solution of the stochastic differential equation

$$X_t = X_0 + \int_0^t b(X_s)\,\mathrm{d}B_s + \int_0^t a(X_s)\,\mathrm{d}s \tag{7.12}$$

driven by a Wiener process B. Here $a : I \to \bar{\mathbb{R}}$, $b : I \to \bar{\mathbb{R}}$ are measurable functions.

(7.13) Definition An I-valued continuous random process X defined on a family of probability spaces $(\Omega,\mathcal{F},\mathbf{P}_x, x \in I)$ as in (1.1) equipped with a filtration $\mathbb{F}$ according to (1.2) is said to be a *solution of equation* (7.12) if

(i) $X \in \mathcal{S}(\mathbb{F})$;
(ii) There exists a Wiener process $(B,\mathbb{F})$ on $(\Omega,\mathcal{F},\mathbf{P}_x, x \in I)$ (cf. (1.11)) such that (7.12) holds for all $t \geq 0$ $\mathbf{P}_x$-a.s. for all $x \in I$.

As in the previous section let $(X,\mathbb{F})$ on $(\Omega,\mathcal{F},\mathbf{P}_x, x \in I)$ be a given strong Markov continuous semimartingale according to (7.1). Recall the decomposition of the set R of regular points into its components: $R = \bigcup_n (a_n, b_n)$.

(7.14) Lemma *For the process*

$$V_t^r = \int_0^t 1_R(X_s)\,\mathrm{d}V_s(X), \qquad t \geq 0,$$

it holds that

$$dV_t^r \ll dt \qquad a.s.$$

if and only if

$$dg \ll m \quad on \quad (K, \mathfrak{B}(K))^2$$

for all compact subintervals $K \subseteq (a_n, b_n)$, $n = 1, 2, \ldots$ *Moreover, in this case we have*

$$V_t^r = \frac{1}{2}\int_0^t \mathbf{1}_{R\cap\{\bar{g}>0\}}(X_s)\frac{dg}{dm}(X_s)\, ds, \qquad t \geq 0,\ a.s.$$

Proof. First assume $dg \ll m$ on $(K, \mathfrak{B}(K))$ for all compact subintervals $K \subseteq (a_n, b_n)$, $n = 1, 2, \ldots$ By setting $\alpha(x) = \frac{dg}{dm}(x)$ from (4.5), (A2.8), (A2.3)(ii) and (5.19) it follows that

$$\begin{aligned} V_t^r &= \frac{1}{2}\int_{R\cap\{\bar{g}>0\}} L^X(t,y)\bar{g}^{-1}(y)\, dg(y) \\ &= \frac{1}{2}\int_{R\cap\{\bar{g}>0\}} L^X(t,y)\bar{g}^{-1}(y)\alpha(y)\, m(dy) \\ &= \frac{1}{2}\int_0^t \mathbf{1}_{R\cap\{\bar{g}>0\}}(X_s)\alpha(X_s)\, ds,\ t \geq 0, \text{ a.s.} \end{aligned}$$

This shows the absolute continuity of V^r as well as the asserted representation.

Now suppose $dV_t^r \ll dt$ a.s. By (1.8) V^r is additive, and, according to (2.23), there exists a $\mathfrak{B}(I)$-measurable function l such that

$$V_t^r = \int_0^t l(X_s)\, ds, \qquad t \geq 0, \text{ a.s.}$$

The remaining part of the proof is analogous to the corresponding part in the proof of (7.8). Let $k : I \to \mathbb{R}$ be a bounded measurable function. Using (4.5), (1.17)(iii) and (5.19) we obtain

$$\begin{aligned} &\int_{\{\bar{g}>0\}\cap R} L^X(t,y)\bar{g}^{-1}(y)k(y)\, dg(y) \\ &\quad = 2\int_0^t \mathbf{1}_{\{\bar{g}>0\}\cap R}(X_s)k(X_s)\, dV_s^r \\ &\quad = 2\int_0^t \mathbf{1}_{\{\bar{g}>0\}\cap R}(X_s)k(X_s)l(X_s)\, ds \\ &\quad = 2\int_{\{\bar{g}>0\}\cap R} L^X(t,y)k(y)l(y)\bar{g}^{-1}(y)\, m(dy),\ t \geq 0, \text{ a.s.} \end{aligned} \tag{7.15}$$

Fix an arbitrary compact subinterval $K \subseteq (a_n, b_n)$ and denote by $K = D_1 \cup D_2$, $D_1 \cap D_2 = \emptyset$, the Hahn decomposition of the signed measure dg on

[2] Here dg denotes the signed measure on $(K, \mathfrak{B}(K))$ induced by g.

$(K, \mathfrak{B}(K))$, i.e. $\mathbf{1}_{D_1} dg$, $-\mathbf{1}_{D_2} dg$ are nonnegative measures. Further, let $B \in \mathfrak{B}(K)$ with $m(B) = 0$ and set $k = \mathbf{1}_{D_1 \cap B} - \mathbf{1}_{D_2 \cap B}$. Then (7.15) yields

$$\int_{\{\bar{g}>0\}\cap R\cap B} L^X(t,y)\bar{g}^{-1}(y)\, \mathrm{d}|g(y)| = 0, \qquad t \geq 0, \text{ a.s.}$$

Using (4.7) and (A2.3)(ii) this leads to

$$|\mathrm{d}g|(B) = 0.$$

But this means $\mathrm{d}g \ll m$ on $(K, \mathfrak{B}(K))$. □

(7.16) Lemma *For $z = a_n \in K_+$ (resp. $z = b_n \in K_-$) it holds that*

$$L^X(\mathrm{d}t, z) \ll \mathrm{d}t \qquad \textit{a.s.}$$

if and only if

$$m(\{z\}) > 0 \quad \textit{provided} \quad z \notin E,\ \bar{g}(z) > 0.$$

Moreover, we then have

$$L^X(t,z) = \frac{\bar{g}(z)}{m(\{z\}\setminus E)} \int_0^t \mathbf{1}_{\{z\}\setminus E}(X_s)\, \mathrm{d}s, \qquad t \geq 0, \textit{ a.s.}$$

Proof. In the case of $z \in E$ or $\bar{g}(z) = 0$, from (1.17)(iv) resp. (4.7) it follows that $L^X(\cdot, z) = 0$ a.s. Hence, there is nothing to show.

Now let $z \notin E$ and $\bar{g}(z) > 0$. First, assume $m(\{z\}) > 0$. Then (5.19) implies

$$L^X(t,z) = \frac{\bar{g}(z)}{m(\{z\})} \int_0^t \mathbf{1}_{\{z\}}(X_s)\, \mathrm{d}s, \qquad t \geq 0, \text{ a.s.}$$

Conversely, if $L^X(\mathrm{d}t, z) \ll \mathrm{d}t$ a.s. then we deduce from (2.23), (1.16), (1.17)(iii) and (5.19) that

$$\begin{aligned} L^X(t,z) &= C \int_0^t \mathbf{1}_{\{z\}}(X_s)\, \mathrm{d}s \\ &= C\, L^X(t,z)\bar{g}^{-1}(z) m(\{z\}) \end{aligned}$$

where $C > 0$ is a constant. Applying (4.7) again, we obtain $m(\{z\}) > 0$. □

(7.17) Lemma *Denote*

$$V_t^+ = \int_0^t \mathbf{1}_{S_+}(X_s)\, \mathrm{d}V_s(X),\ V_t^- = \int_0^t \mathbf{1}_{S_-}(X_s)\, \mathrm{d}V_s(X), \qquad t \geq 0.$$

Then $\mathrm{d}V_t^+ \ll \mathrm{d}t$ (resp. $\mathrm{d}V_t^+ \ll \mathrm{d}t$) holds a.s. if and only if

$$h > 0 \quad \textit{on} \quad S_+ \ (\textit{resp. } S_-) \quad \textit{Lebesgue-almost everywhere,}$$

where h is defined by (7.2). Moreover, then we have

$$V_t^+ = \int_0^t \mathbf{1}_{S_+}(X_s)h^{-1}(X_s)\,\mathrm{d}s \quad \left(\textit{resp. } V_t^- = \int_0^t \mathbf{1}_{S_-}(X_s)h^{-1}(X_s)\,\mathrm{d}s\right),$$

$$t \geq 0, \textit{ a.s.}$$

Proof. By symmetry, we only consider the process V^+. First assume $h > 0$ on S_+ Lebesgue-almost everywhere. Because of $\ell(N) = 0$ for N in (7.2) and using (4.13), (2.14) we get

$$\begin{aligned} V_t^+ &= \int_{[X_0, \max_{s\leq t} X_s]\cap S_+} \mathbf{1}_{N^c}(y)\,\mathrm{d}y \\ &= \int_{[X_0, \max_{s\leq t} X_s]\cap S_+} h^{-1}(y)\mathbf{1}_{N^c}(y)h(y)\,\mathrm{d}y \\ &= \int_{[X_0, \max_{s\leq t} X_s]\cap S_+} h^{-1}(y)\mathbf{1}_{N^c}(y)\,m(\mathrm{d}y), \quad t \geq 0, \text{ a.s.} \end{aligned}$$

By (5.19) this implies

$$V_t^+ = \int_0^t \mathbf{1}_{S_+}(X_s)(\mathbf{1}_{N^c}h^{-1})(X_s)\,\mathrm{d}s, \qquad t \geq 0, \text{ a.s.}$$

Taking into account (1.28) and (7.3) this is the desired representation.

Now, let $\mathrm{d}V_t^+ \ll \mathrm{d}t$ a.s. Applying (2.23) again, there exists a $\mathfrak{B}^u(I)$-measurable function l such that

$$V_t^+ = \int_0^t \mathbf{1}_{S_+}(X_s)l(X_s)\,\mathrm{d}s, \qquad t \geq 0, \text{ a.s.}$$

For each $A \in \mathfrak{B}(I)$, from (5.19), (7.2) we deduce

$$\begin{aligned} \int_0^t \mathbf{1}_A(X_s)\,\mathrm{d}V_s^+ &= \int_0^t \mathbf{1}_{S_+\cap A}(X_s)l(X_s)\,\mathrm{d}s \\ &= \int_{[X_0, \max_{s\leq t} X_s]\cap S_+\cap A} l(y)h(y)\,\mathrm{d}y \\ &\quad + \int_{[X_0, \max_{s\leq t} X_s]\cap S_+\cap A} l(y)\,n(\mathrm{d}y) \end{aligned} \tag{7.18}$$

for all $t \geq 0$ a.s. On the other hand, (4.13), (2.14) yield

$$\int_0^t \mathbf{1}_A(X_s)\,\mathrm{d}V_s^+ = \int_{[X_0, \max_{s\leq t} X_s]\cap S_+} \mathbf{1}_A(y)\,\mathrm{d}y. \tag{7.19}$$

Combining (7.18) and (7.19), for $A = \{h = 0\} \cap S_+ \setminus N$ we obtain

$$\ell([X_0, \max_{s\leq t} X_s] \cap A) = 0, \qquad t \geq 0, \text{ a.s.}$$

Hence, for all $x \notin K_-$ there exists an $\varepsilon > 0$ such that

$$\ell(A \cap [x, x+\varepsilon)) = 0. \tag{7.20}$$

Because of K_- is closed to the left-hand side (see (2.5)) we may represent $I \setminus K_-$ as a union of an at most countable number of intervals of the form $[a, b)$ or (a, b). For each such interval, (7.20) implies

$$\ell(A \cap [a, b)) = 0 \quad \text{or} \quad \ell(A \cap (a, b)) = 0,$$

respectively. This leads to

$$\ell(A \setminus K_-) = \ell(A) = 0,$$

i.e. $h > 0$ on S_+ Lebesgue-almost everywhere. □

Combining (4.3) and (7.8), (7.14), (7.16), (7.17) we now obtain the main result of this section.

(7.21) Theorem *Let $(X, \mathbb{F})$ on $(\Omega, \mathcal{F}, \mathbf{P}_x, x \in I)$ be a continuous strong Markov semimartingale with speed measure m associated with a scale function p whose Lebesgue decomposition is given by (7.2), (7.3). Then there exist $\mathfrak{B}^u(I)$-measurable functions $a : I \to \bar{\mathbb{R}}$ and $b : I \to \bar{\mathbb{R}}$ such that $(X, \mathbb{F})$ is a solution of the stochastic differential equation*

$$X_t = X_0 + \int_0^t b(X_s)\,\mathrm{d}B_s + \int_0^t a(X_s)\,\mathrm{d}s$$

if and only if the following conditions are satisfied:

(i) *$h > 0$ on $\{x \in R : \bar{g}(x) > 0\} \cup S_+ \cup S_-$ Lebesgue-almost everywhere,*
(ii) *$\mathrm{d}g \ll m$ on $(K, \mathfrak{B}(K))$ for all compact subintervals $K \subseteq R$,*
(iii) *for all $z \in \{a_n : a_n \in K_+ \setminus E\} \cup \{b_n : b_n \in K_- \setminus E\}$, it holds that $m(\{z\}) > 0$ if $\bar{g}(z) > 0$.*

Moreover we can set

$$\begin{aligned} b &= \sqrt{\mathbf{1}_R h^{-1} \bar{g}}, \\ a &= \frac{1}{2}\mathbf{1}_{R \cap \{\bar{g} > 0\}} \frac{\mathrm{d}g}{\mathrm{d}m} + (\mathbf{1}_{S_+} - \mathbf{1}_{S_-}) h^{-1} \\ &\quad + \sum_{z \in \{a_n : a_n \in K_+ \setminus E\}} \frac{\bar{g}(z)}{m(\{z\})} \mathbf{1}_{\{z\}} \\ &\quad - \sum_{z \in \{b_n : b_n \in K_- \setminus E\}} \frac{\bar{g}(z)}{m(\{z\})} \mathbf{1}_{\{z\}}. \end{aligned}$$

VII.3 Fundamental Examples of Non-regular Diffusions

The last Theorem (7.21) turns out to be an important link between the construction method of Chapter VI and the decomposition of continuous strong Markov semimartingales given in Chapter IV. Indeed, it allows us to construct strong Markov Itô processes[3] as examples for diffusions in all typical cases of non-regularity. To be more precise, to get a continuous strong Markov process possessing a certain behaviour in the state space one only has to specify the corresponding sets R, K_+, K_-, and, for properly chosen speed measure and scale, the construction method of Chapter VI and the results of Section VII.2 give us an Itô process with the desired properties. The coefficients of the stochastic differential equation solved by the obtained process contain some information about R, K_+, K_-, as well as about the scale function and the speed measure we have chosen for the construction.

As one can see in the examples below, the drift and diffusion functions of the stochastic differential equation can be highly irregular. Those functions are usually outside the range of known theorems on the existence of solutions of stochastic differential equations.

Finally we remark that Itô processes with a non-regular behaviour in the state space were constructed in [1],[2] using similar methods. Now we come to the announced examples. At first we treat the case where the set of regular points is empty.

(7.22) Example Let $(0,1) = I = K_+ \cup K_-$. We suppose that I will be connected with respect to the continuous strong Markov process we are going to construct. Because of the remarks after (2.4) this assumption is not a vital restriction. Applying Definition (2.10) of the sets S_+, S_- we immediately obtain that I is the disjoint union

$$(0,1) = S_+ \cup S_- \cup E.$$

Assume for a moment that we have already constructed a continuous strong Markov process with the decomposition $S_+ \cup S_- \cup E$ of its state space $I = (0,1)$ and such that I is connected. We show that S_+, S_- and E necessarily admit a very simple structure: S_+, S_- are open intervals and E is at most a one-point set.

To start with, we show that S_+ and S_- are open subsets of $(0,1)$. We only consider S_+, the result for S_- is proved analogously.

Let $x \in S_+$ be fixed and assume that x is not an inner point of S_+. Hence, for all $\varepsilon > 0$, $K_- \cap (x-\varepsilon, x+\varepsilon) \neq \emptyset$, and, we can choose a sequence $(x_n) \subseteq K_-$ converging to x. But, recalling (2.5)(ii), K_- is a closed subset of $(0,1)$. Therefore, it holds that x belongs to K_- which is a contradiction. Summing up, all points of S_+ are inner points proving that S_+ is open.

Now we decompose S_+ and S_- into their components

[3] Solutions of (7.12).

$$S_+ = \bigcup_n (c_n, d_n) \quad \text{and} \quad S_- = \bigcup_n (\tilde{c}_n, \tilde{d}_n)$$

and consider a fixed component $(c_n, d_n) \subseteq S_+$. If $c_n > 0$ then, applying (2.5)(ii) again, we have that $c_n \in K_+ \setminus E = S_+$ which is a contradiction. Therefore, $c_n = 0$.

Summing up we get that $c_n = 0$ for all components $(c_n, d_n) \subseteq S_+$. This yields that, if $K_+ \neq \emptyset$, then there exists a $c \in (0,1]$ such that

$$S_+ = (0, c).$$

Clearly, if $K_- = \emptyset$ then $c = 1$. Analogously one can prove that

$$S_- = (c, 1).$$

We now consider the case $K_+, K_- \neq \emptyset$ and construct a continuous strong Markov process with state space

$$(0,1) = S_+ \cup S_- \cup E = (0,c) \cup (c,0) \cup \{c\}$$

for some $c \in (0,1)$. According to (6.3) we choose a function $p : (0,1) \to \mathbb{R}$, for example $p(x) = x$, and specify a p-admissible measure m on $((0,1), \mathfrak{B}((0,1)))$, that is a measure with the properties (see (6.4)):

$$\begin{gathered}
\forall G \subseteq (0,c) \cup (c,1),\ G \text{ open} : m(G) > 0, \\
\forall x \in (0,c) \cup (c,1) : m(\{x\}) = 0, \\
\forall x \in (0,c) \exists \varepsilon > 0 : m([x, (x+\varepsilon) \wedge c)) < \infty, \\
\forall x \in (c,1) \exists \varepsilon > 0 : m(((x-\varepsilon) \vee c, x]) < \infty, \\
\forall \varepsilon > 0 : m(((c-\varepsilon) \vee 0, (c+\varepsilon) \wedge 1)) = \infty.
\end{gathered}$$

Obviously, the measure

$$m(\mathrm{d}x) = h(x)\,\mathrm{d}x$$

with

$$h(x) = \begin{cases} (c-x)^{-1} & : \ x \in (0,c) \\ \infty & : \ x = c \\ (x-c)^{-1} & : \ x \in (c,1) \end{cases}$$

possesses the properties above, and, consequently, by Theorem (6.48) there exists a continuous strong Markov process $(X, \mathbb{F})$ on a probability space $(\Omega, \mathcal{F}, \mathbf{P}_x, x \in (0,1))$ admitting $\emptyset$, $(0,c]$, $[c,1)$ as the set of regular, right singular, left singular points, respectively, and $m|_{(0,1)\setminus\{c\}}$ as speed measure associated with p. On the other hand, $(X, \mathbb{F})$ also satisfies the conditions of Theorem (7.21) because of $R = \emptyset$ and $h > 0$ on $(0,c) \cup (c,1)$. Therefore, $(X, \mathbb{F})$ is an Itô process solving the deterministic differential equation

$$X_t = X_0 + \int_0^t (c - X_s)\,\mathrm{d}s.$$

The flow of solutions of the last equation is well-known and its behaviour can be viewed as the "typical" behaviour of a continuous strong Markov process having no regular points in the state space.

(7.23) Example In this example we assume that $K_- = \emptyset$ and study the case of an isolated right singular point.

Set $I = \mathbb{R}$, $R = (-\infty, c) \cup (c, \infty)$, $K_+ = \{c\}$, $K_- = \emptyset$ and define

$$p(x) = x, \ x \in \mathbb{R},$$

as well as

$$m = \ell + \delta_c.$$

Then m is a p-admissible measure on $(\mathbb{R}, \mathfrak{B}(\mathbb{R}))$ (see (6.4)). Again, Theorem (6.48) yields the existence of a continuous strong Markov process $(X, \mathbb{F})$ on a probability space $(\Omega, \mathcal{F}, \mathbf{P}_x, x \in \mathbb{R})$ admitting $(-\infty, c) \cup (c, \infty)$, $\{c\}$, $\emptyset$ as the set of regular, right singular, left singular points, respectively, and m as speed measure associated with p. Further we compute[4]

$$h(x) = \begin{cases} 1 & : \ x \in (-\infty, c) \cup (c, \infty) \\ \infty & : \ x = c \end{cases},$$

$$g(x) = 1, \ x \in (-\infty, c) \cup (c, \infty),$$

$$\bar{g}(x) = 1, \ x \in \mathbb{R},$$

and all conditions of Theorem (7.21) are obviously satisfied. Therefore, the above constructed process $(X, \mathbb{F})$ is an Itô process solving the stochastic differential equation

$$X_t = X_0 + \int_0^t \mathbf{1}_{\mathbb{R}\setminus\{c\}}(X_s) \, \mathrm{d}B_s + \int_0^t \mathbf{1}_{\{c\}}(X_s) \, \mathrm{d}s.$$

We remark that the solution of this equation is not unique in law. For example, the Wiener process $(B, \mathbb{F})$ driving the equation is also a solution with a law different from $(X, \mathbb{F})$ because it is a regular diffusion. But all solutions $(X, \mathbb{F})$ satisfying

$$X_t \geq c, \qquad t \geq \mathrm{D}_{[c,\infty)}(X), \text{ a.s.},$$

admit the same law, namely the law of the process we constructed above. The proof of this uniqueness-result is far from the topic of these lecture notes and therefore omitted. We refer to [2].

(7.24) Example Now, let us investigate the case of an isolated absorbing point. Again, we set $I = \mathbb{R}$, $R = (-\infty, c) \cup (c, \infty)$, $K_+ = \{c\}$ as in (7.23), but change $K_- = \emptyset$ into $K_- = \{c\}$. Choose again

$$p(x) = x, \ x \in \mathbb{R}.$$

Then the measure m defined in (7.23) is not a p-admissible measure since (6.4)(vi) is violated. However,

[4] See (7.2),(7.3) for the definition of h.

$$m = \ell + \infty \cdot \delta_c$$

turns out to be a p-admissible measure on $(\mathbb{R}, \mathfrak{B}(\mathbb{R}))$, and, using Theorem (6.48), we construct a continuous strong Markov process $(X, \mathbb{F})$ on a probability space $(\Omega, \mathcal{F}, \mathbf{P}_x, x \in \mathbb{R})$ admitting $(-\infty, c) \cup (c, \infty)$, $\{c\}$, $\{c\}$ as the set of regular, right singular, left singular points, respectively, and $m|_{\mathbb{R}\setminus\{c\}}$ as speed measure associated with p. Clearly, we obtain

$$h(x) = \begin{cases} 1 & : \; x \in (-\infty, c) \cup (c, \infty) \\ \infty & : \; x = c \end{cases},$$

$$g(x) = 1, \; x \in (-\infty, c) \cup (c, \infty),$$

$$\bar{g}(x) = 1, \; x \in \mathbb{R}.$$

We remark that these functions are equal to the corresponding ones in Example (7.23). Again, Theorem (7.21) can be applied and $(X, \mathbb{F})$ is an Itô process solving the stochastic differential equation

$$X_t = X_0 + \int_0^t \mathbf{1}_{\mathbb{R}\setminus\{c\}}(X_s)\, \mathrm{d}B_s.$$

As in the previous example, the solution to this stochastic differential equation is not unique in law, see [24] for a characterization of all possible solutions. It is an easy consequence of [23], Theorem (5.18), that all solutions $(X, \mathbb{F})$ of the last equation additionally satisfying

$$X_t = X_{t \wedge D_{\{c\}}}(X), \qquad t \geq 0, \text{ a.s.},$$

admit the same law as the process we constructed above.

The following two examples investigate the situation where $R \neq \emptyset$ and accumulation points of K_+ appear.

(7.25) Example Set $I = \mathbb{R}$, $R = (-\infty, 0) \cup \bigcup_{n=1}^{\infty} (x_n, x_{n-1}) \cup (x_0, \infty)$ with $x_n < x_0$, $x_n \downarrow 0$, $K_+ = \{0\} \cup \{x_n : n = 0, ..., \infty\}$, $K_- = \emptyset$. Hence, $0 \in S_+$ because $K_- = \emptyset$. Remark that the limit point of $(x_n)_{n=0}^{\infty}$ is necessarily in S_+ if I is assumed to be connected with respect to the continuous strong Markov process we want to construct. Further we define

$$p(x) = x, \; x \in \mathbb{R},$$

and

$$m = \ell + \sum_{n=0}^{\infty} \delta_{x_n}.$$

In order to show that m is a p-admissible measure on $(\mathbb{R}, \mathfrak{B}(\mathbb{R}))$ we only verify (6.4)(iv) for $x = 0$. The other conditions of (6.4) are obviously satisfied. Fix n_0 such that $x_{n_0-1} \leq 1$, set $\varepsilon = x_{n_0-1}$ and compute

$$\sum_n \int_{[a_n,b_n)\cap[0,\varepsilon)} (p(b_n \wedge \varepsilon) - p(y))\, m(\mathrm{d}y) \,+\, m([0,\varepsilon)\cap S_+)$$
$$= \sum_{n=n_0}^{\infty} \left[\int_{x_n}^{x_{n-1}} (x_{n-1} - y)\,\mathrm{d}y \,+\, (x_{n-1} - x_n) \right]$$
$$\leq 2 \sum_{n=n_0}^{\infty} (x_{n-1} - x_n) \,=\, 2\varepsilon \,<\, \infty.$$

Hence, from Theorem (6.48) it follows that there exists a continuous strong Markov process $(X, \mathbb{F})$ on a probability space $(\Omega, \mathcal{F}, \mathbf{P}_x, x \in \mathbb{R})$ admitting $(-\infty, 0) \cup \bigcup_{n=1}^{\infty} (x_n, x_{n-1}) \cup (x_0, \infty)$, $\{0\} \cup \{x_n : n = 0, ..., \infty\}$, $\emptyset$ as the set of regular, right singular, left singular points, respectively, and m as the speed measure associated with p. For h defined as in (7.2),(7.3) we have that

$$h(x) \;=\; \begin{cases} 1 & : \; x \in \mathbb{R} \setminus (x_n)_{n=0}^{\infty} \\ \infty & : \; x = x_n,\ n = 0, 1, 2, \ldots \end{cases},$$

$$g(x) \;=\; 1,\ x \in \mathbb{R} \setminus (\{0\} \cup \{x_n : n = 0, ..., \infty\}),$$

$$\bar{g}(x) \;=\; 1,\ x \in \mathbb{R} \setminus \{0\},$$

and, Theorem (7.21) yields that $(X, \mathbb{F})$ is an Itô process solving the stochastic differential equation

$$X_t \,=\, X_0 \,+\, \int_0^t \mathbf{1}_{\mathbb{R}\setminus(\{0\}\cup(x_n)_{n=0}^{\infty})}(X_s)\,\mathrm{d}B_s \,+\, \int_0^t \mathbf{1}_{\{0\}}(X_s)\,\mathrm{d}s$$
$$+ \int_0^t \sum_{n=0}^{\infty} \mathbf{1}_{\{x_n\}}(X_s)\,\mathrm{d}s.$$

But, combining $m(\{0\}) = 0$, $0 \in S_+$ and the occupation time formula (5.27), we obtain

(7.26) $$\int_0^t \mathbf{1}_{\{0\}}(X_s)\,\mathrm{d}s \,=\, 0$$

for all $t \geq 0$ a.s. That is, the process X has no occupation time in the right singular point 0. The right singularity of X at 0 is forced by the right singular points (x_n) in every right neighbourhood of 0. For isolated right singular points such a phenomenon is not possible. Further, because of (7.26), $(X, \mathbb{F})$ also solves the stochastic differential equation

$$X_t \,=\, X_0 \,+\, \int_0^t \mathbf{1}_{\mathbb{R}\setminus(x_n)_{n=0}^{\infty}}(X_s)\,\mathrm{d}B_s \,+\, \int_0^t \sum_{n=0}^{\infty} \mathbf{1}_{\{x_n\}}(X_s)\,\mathrm{d}s$$

whose diffusion coefficient does not vanish at the point 0. As a consequence, all solutions $(X, \mathbb{F})$ of this stochastic differential equation additionally satisfying

$$X_t \geq x_n, \qquad t \geq \mathrm{D}_{x_n}(X), \text{ a.s.,}$$

$n = 0, 1, 2, ...,$ have the same law; that is the system

$$\begin{cases} X_t &= X_0 + \int_0^t \mathbf{1}_{\mathbb{R}\setminus(x_n)_{n=0}^\infty}(X_s)\,\mathrm{d}B_s + \int_0^t \sum_{n=0}^\infty \mathbf{1}_{\{x_n\}}(X_s)\,\mathrm{d}s \\ X_t &\geq x_n,\, t \geq \mathrm{D}_{x_n}(X),\, n = 0, 1, 2, ... \end{cases}$$

is characteristic for a continuous strong Markov process admitting $(-\infty, 0) \cup \bigcup_{n=1}^\infty (x_n, x_{n-1}) \cup (x_0, \infty)$, $\{0\} \cup \{x_n : n = 0, ..., \infty\}$, $\emptyset$ as the set of regular, right singular, left singular points, respectively, and $\ell + \sum_{n=0}^\infty \delta_{x_n}$ as speed measure associated with $p(x) = x$. As in Example (7.23) we do not prove this uniqueness result here, the reader is refered to [2].

(7.27) Example Set $I = (-\infty, 1)$, $R = (-\infty, x_0) \cup \bigcup_{n=0}^\infty (x_n, x_{n+1})$ with $x_0 < x_n$, $x_n \uparrow 1$, $K_+ = \{x_n : n = 0, ..., \infty\}$, $K_- = \emptyset$. In this case we have $S_+ = \emptyset$. Define

$$p(x) = x,\; x \in (-\infty, 1).$$

Looking for a p-admissible measure m on $((-\infty, 1), \mathfrak{B}((-\infty, 1)))$, the critical condition is (6.4)(v). But the measure

$$m = \ell|_{(-\infty,1)} + \sum_{n=0}^\infty \frac{1}{x_{n+1} - x_n} \delta_{x_n}$$

on $((-\infty, 1), \mathfrak{B}((-\infty, 1)))$ is p-admissible. Indeed, (6.4)(v) is satisfied since for all n it holds that

$$\begin{aligned} &\sum_{k=n}^\infty \int_{[x_k, x_{k+1})} (p(x_{k+1}) - p(y))\, m(\mathrm{d}y) \\ &\quad = \sum_{k=n}^\infty \left[\int_{x_k}^{x_{k+1}} (x_{k+1} - y)\,\mathrm{d}y + (x_{k+1} - x_k) \frac{1}{x_{k+1} - x_k} \right] \\ &\quad \geq \sum_{k=n}^\infty 1 = \infty. \end{aligned}$$

The remaining properties of (6.4) are evidently satisfied. Applying Theorem (6.48) again, there is a continuous strong Markov process $(X, \mathbb{F})$ on a probability space $(\Omega, \mathcal{F}, \mathbf{P}_x, x \in (-\infty, 1))$ admitting $(-\infty, x_0) \cup \bigcup_{n=0}^\infty (x_n, x_{n+1})$, $(x_n)_{n=0}^\infty$, $\emptyset$ as the set of regular, right singular, left singular points, respectively, and m as speed measure associated with p. The point 1 is not attainable for the process X although in each left-hand neighbourhood of it there is a right singular point. This is a consequence of the condition (6.4)(v). Indeed, (6.4)(v) can be viewed as a condition ensuring that the corresponding continuous strong Markov process does not "leave" its state interval I (see the proof of (5.34)(vi)). Remark that for regular strong Markov processes X the relationship between (6.4)(v) and the behaviour of X in I is well-known (cf. [34] for example).

Finally, the conditions of Theorem (7.21) are also satisfied and (cf. (7.2),(7.3))

$$
\begin{aligned}
h(x) &= \begin{cases} 1 & : \quad x \in (-\infty, 1) \setminus (x_n)_{n=0}^{\infty} \\ \infty & : \quad x = x_n,\ n = 0, 1, 2, \ldots \end{cases}, \\
g(x) &= 1,\ x \in (-\infty, 1) \setminus (x_n)_{n=0}^{\infty}, \\
\bar{g}(x) &= 1,\ x \in (-\infty, 1).
\end{aligned}
$$

Consequently, $(X, \mathbb{F})$ is an Itô process solving the stochastic differential equation

$$X_t = X_0 + \int_0^t \mathbf{1}_{(-\infty,1)\setminus(x_n)_{n=0}^{\infty}}(X_s)\, \mathrm{d}B_s + \int_0^t \sum_{n=0}^{\infty} (x_{n+1} - x_n) \mathbf{1}_{\{x_n\}}(X_s)\, \mathrm{d}s.$$

Again one can show that all $(-\infty, 1)$-valued solutions $(X, \mathbb{F})$ of the last equation additionally satisfying

$$X_t \geq x_n, \qquad t \geq \mathrm{D}_{x_n}(X), \text{ a.s.},$$

$n = 0, 1, 2, \ldots$, admit the same law, namely the law of the process we constructed above.

Appendix 1

Semimartingale Functions and the Behaviour of Integral Functionals of the Wiener Process

In this Appendix we provide some results which are important for our considerations in Chapter IV and VI. On the other hand, the results below are also of independent interest.

Reflected Wiener Process

Let $(\Omega, \mathcal{F}, \mathbf{P})$ be a complete probability space equipped with a right-continuous filtration $\mathbb{F}$ satisfying (1.1) and (1.2)(iii)[1]. Fix a non-empty interval $[a,b]$, $a,b \in \bar{\mathbb{R}}$, as well as an $\mathbb{F}$-stopping time T.

(A1.1) Definition A continuous semimartingale $W \in \mathcal{S}(\mathbb{F})$ with decomposition

$$W = W_0 + M + V$$

is said to be a *Wiener process reflected in* $[a,b]$ *and stopped at* T if

(i) $$W_0 \in [a,b],$$

(ii) $$\langle M \rangle_t = t \wedge T, \qquad t \geq 0, \text{ a.s.},$$

(iii) $$V_t = L^W(t,a) - L^W(t,b), \qquad t \geq 0, \text{ a.s.}$$

In the case of $T = \infty$ we also call $(W, \mathbb{F})$ a *Wiener process reflected in* $[a,b]$.

(A1.2) Remark (i) Computing $(W-a)^+$ if $a > -\infty$ (resp. $(W-b)^+$ if $b < -\infty$) by the generalized Itô-formula (1.14) one easily sees that for a Wiener process reflected in $[a,b]$ and stopped at T it holds that $W_t \in [a,b]$, $t \geq 0$, a.s.

(ii) In view of our definition of $L^W(\cdot, \pm\infty) = 0$ (see page 6), a Wiener process reflected in $[-\infty, +\infty]$ is nothing else but a Wiener process.

---

[1] By $I = [0,0]$, $\mathbf{P}_0 = \mathbf{P}$ this setup is embedded in the setting (1.1).

(A1.3) The existence of a Wiener process reflected in $[a, b]$ is easily seen. We outline a possible construction.

Let $(W, \mathbb{G})$ be a Wiener process on a probability space $(\Omega, \mathcal{G}, \mathbf{P})$. We put

$$X_t = (W_t \vee a) \wedge b.$$

Obviously $(X, \mathbb{G})$ is a semimartingale. Applying the generalized Itô-formula (1.14) we get

$$X_t = X_0 + \int_0^t \mathbf{1}_{[a,b)}(W_s) \mathrm{d}W_s + \frac{1}{2} L_-^W(t, a) - \frac{1}{2} L_-^W(t, b).$$

Using (1.15) it is easy to verify that

$$L^X(t, a) = \frac{1}{2} L_-^W(t, a), \; L^X(t, b) = \frac{1}{2} L_-^W(t, b), \; L^X(t, x) = L_-^W(t, x), \; x \in (a, b).$$

Now let

$$\begin{aligned} A_t &= \langle X \rangle_t = \int_0^t \mathbf{1}_{[a,b)}(W_s) \mathrm{d}s, \\ T_t &= \inf\{s \geq 0 : A_s > t\}. \end{aligned}$$

Then $A_t < \infty$, $t \geq 0$, $A_\infty = \infty$ **P**-a.s. (cf. [35], Ch. 3.6) and thus

$$T_t < \infty, \; t \geq 0, \; T_\infty = \infty, \; \mathbf{P} - a.s.$$

The processes $\int_0^t \mathbf{1}_{[a,b)}(W_s) \mathrm{d}W_s$, $L^X(t, a)$ and $L^X(t, b)$ are constant on the intervals $[T_{t-}, T_t]$, $T_{0-} = 0$. Therefore, by Proposition (10.17) in [32] we conclude that the time-changed process $Y_t = X_{T_t}$, $\mathcal{F}_t = \mathcal{G}_{T_t}$ is a continuous semimartingale with

$$\begin{aligned} Y_t &= Y_0 + \int_0^{T_t} \mathbf{1}_{[a,b)}(W_s) \mathrm{d}W_s + L^X(T_t, a) - L^X(T_t, b) \\ &= Y_0 + B_t + L^Y(t, a) - L^Y(t, b), \end{aligned}$$

where $(B, \mathbb{F})$ is a Wiener process. Obviously $(Y, \mathbb{F})$ is a Wiener process reflected in $[a, b]$.

(A1.4) Lemma *Let $(W, \mathbb{F})$ be a Wiener process reflected in $[a, b]$ and stopped at T. Then there exists a Wiener process $(\widetilde{W}, \widetilde{\mathbb{F}})$ reflected in $[a, b]$, possibly defined on an extension $(\widetilde{\Omega}, \widetilde{\mathcal{F}}, \widetilde{\mathbf{P}})$, $\widetilde{\mathbb{F}}$ of $(\Omega, \mathcal{F}, \mathbf{P})$, $\mathbb{F}$ (see (1.25)), such that $W = \widetilde{W}^T$.*

Proof. Without loss of generality on $(\Omega, \mathcal{F}, \mathbf{P})$ we may assume the existence of an $\mathbb{F}$-adapted Wiener process B, otherwise, we extend the stochastic basis according to (1.25).

For $t \geq 0$ define

$$\begin{aligned} X_t &= W_t + (B_t - B_{t\wedge T}) \\ Z_t &= (X_t \vee a) \wedge b, \\ A_t &= \langle Z \rangle_t = \int_0^t \mathbf{1}_{[a,b)}(X_s)\, ds, \\ T_t &= \inf\{s \geq 0 : A_s > t\}, \\ \widetilde{W}_t &= Z_{T_t}, \\ \widetilde{\mathcal{F}}_t &= \mathcal{F}_{T_t}. \end{aligned}$$

Obviously, for $t < T$ it holds that $Z_t = W_t$, $A_t = t$ and $T_t = t$ (see (A1.2)(i)), consequently, $W_t = \widetilde{W}_{t\wedge T}$. Finally as in (A1.3) one shows that $\widetilde{W}$ is a Wiener process reflected in $[a, b]$. □

(A1.5) Proposition *Let $Y \in \mathcal{S}(\mathbb{F})$ be a continuous semimartingale with decomposition*

$$Y = Y_0 + M + L^Y(t,a) - L^Y(t,b); \; M \in \mathcal{M}(\mathbb{F}),$$

satisfying $Y_0 \in [a,b]$, $a < b$, $a, b \in \bar{\mathbb{R}}$. For $T_t = \inf\{s \geq 0 : \langle Y \rangle_s > t\}$ denote $W = Y \circ T$, $\mathbb{G} = \mathbb{F} \circ T$. Then $(W, \mathbb{G})$ is a Wiener process reflected in $[a, b]$ and stopped at $\langle Y \rangle_\infty$.

The proof of this assertion is straightforward, we refer to the proof of Proposition 3 in [42].

Let $(W, \mathbb{F})$ be a Wiener process reflected in $[a, b]$ and stopped at T. We investigate the question for which functions $f : [a, b] \to \mathbb{R}$ it holds that $f(W) \in \mathcal{S}(\mathbb{F})$, that is, f is a semimartingale function for $(W, \mathbb{F})$. As is well-known, for the standard Wiener process the set of semimartingale functions coincides with the set of all functions that can be represented as a difference of convex functions (see [48]). For a wide class of Markov processes the corresponding semimartingale functions have been completely characterized in [11].

A real function f defined on an interval I is said to be *convex on* I, if, on each compact interval $K \subseteq I$, the second derivative of f in the sense of distributions can be identified with a finite measure on $(K, \mathfrak{B}(K))$.

(A1.6) Proposition *Let $(W, \mathbb{F})$ be a Wiener process reflected in $[a, b]$ and let $f : [a, b] \to \mathbb{R}$ be a measurable function.*

(i) The function f is a semimartingale function for $(W, \mathbb{F})$ if and only if f can be represented as a difference of convex functions on $[a, b]$.

(ii) Let $(c,d) \subseteq [a,b]$ *and* $W_0 \in (c,d)$. *If* f *is a semimartingale function for* $(W^{D_{(c,d)^c}}, \mathbb{F})$ *then* $f|_{(c,d)}$ *can be represented as a difference of convex functions on* (c,d).

(iii) Let $a < c < b$ *and* $W_0 \in [a,c)$. *If* f *is a semimartingale function for* $(W^{D_c}, \mathbb{F})$ *then* $f|_{[a,c)}$ *can be represented as a difference of convex functions on* $[a,c)$.

For the proof see [11], ch. 5 or [48], for example.

The Behaviour of Integral Functionals of the Wiener Process

The following useful lemma is a slight generalization of a result of Jeulin [33], Xue [50].

(A1.7) Lemma *Let* $B \in \mathfrak{B}(\mathbb{R})$ *be fixed and* $(L_y)_{y\in B}$ *be a family of nonnegative random variables on a probability space* $(\Omega, \mathcal{F}, \mathbf{P})$ *with* $\mathbf{P}(\{L_y = 0\}) = 0$, $y \in B$. *Assume that the mapping* $(y,\omega) \mapsto L_y(\omega)$ *is* $\mathfrak{B}(B)\otimes\mathcal{F}$*-measurable. Further, assume that there exists a measurable function* $\phi : B \to (0,\infty)$ *such that* $L_y/\phi(y)$ *possesses the same distribution as a certain integrable random variable* ξ, *for each* $y \in B$. *Let* m *be a measure on* $(B, \mathfrak{B}(B))$ *such that* $\phi\,\mathrm{d}m$ *is* σ*-finite. Then the following assertions are equivalent:*

(i) $$\mathbf{P}\left(\left\{\int_B L_y\, m(\mathrm{d}y) < \infty\right\}\right) > 0.$$

(ii) $$\mathbf{P}\left(\left\{\int_B L_y\, m(\mathrm{d}y) < \infty\right\}\right) = 1.$$

(iii) $$\int_B \phi(y)\, m(\mathrm{d}y) < \infty.$$

Proof. (Compare [50], Lemma 2.) Using the Theorem of Fubini the implication (iii) $\Longrightarrow$ (ii) immediately follows from

$$\mathbf{E}\int_B L_y\, m(\mathrm{d}y) = \mathbf{E}\int_B L_y/\phi(y)\,\phi(y)\, m(\mathrm{d}y) = \mathbf{E}\xi\int_B \phi(y)\, m(\mathrm{d}y) < \infty.$$

Now, we assume (i) and show (iii). First, (i) yields that there exists an $N > 0$ such that

$$\mathbf{P}\left(\left\{\int_B L_y\, m(\mathrm{d}y) \le N\right\}\right) > 0.$$

For $A = \left\{\int_B L_y\, m(\mathrm{d}y) \le N\right\}$, applying the Theorem of Fubini again, we obtain

$$\begin{aligned} N &\ge \mathbf{E}\left(\mathbf{1}_A \int_B L_y\, m(\mathrm{d}y)\right) \\ &= \int_B \mathbf{E}\left(\mathbf{1}_A L_y/\phi(y)\right)\phi(y)\, m(\mathrm{d}y) \end{aligned}$$

$$\begin{aligned} &= \int_B \int_0^\infty \mathbf{P}(A \cap \{L_y/\phi(y) > u\})\, du\, \phi(y)\, m(dy) \\ &\geq \int_B \int_0^\infty (\mathbf{P}(A) - \mathbf{P}(\{L_y/\phi(y) \leq u\}))^+\, du\, \phi(y)\, m(dy) \\ &= \int_B \phi(y)\, m(dy) \int_0^\infty (\mathbf{P}(A) - \mathbf{P}(\{\xi \leq u\}))^+\, du. \end{aligned}$$

But, $\mathbf{P}(\{\xi = 0\}) = \mathbf{P}(\{L_y = 0\}) = 0$ implies

$$\int_0^\infty (\mathbf{P}(A) - \mathbf{P}(\{\xi \leq u\}))^+\, du > 0$$

leading to (iii). □

Using the lemma above, we are able to derive the following zero-one law for the convergence of certain functionals of the Wiener process.

(A1.8) Proposition *Let* $(W, \mathbb{G})$ *on* $(\Omega, \mathcal{F}, \mathbf{P})$ *be a Wiener process with* $W_0 = x_0$. *Moreover, let* $c_n \in \mathbb{R}$, $n = 1, 2, \ldots$, *and let us be given disjoint sets* $I_n \in \mathfrak{B}(\mathbb{R})$ *such that, for each* n, $I_n \subseteq [x_0, c_n)$ *or* $I_n \subseteq (c_n, x_0]$. *We set* $B = \bigcup_n I_n$. *For a measure* m *on* $(B, \mathfrak{B}(B))$ *we consider the functional*

$$A = \sum_n \int_{I_n} L^W(\mathrm{D}_{c_n}, y)\, m(dy).$$

Then the following assertions are equivalent:

(i) $\mathbf{P}(\{A < \infty\}) > 0.$

(ii) $\mathbf{P}(\{A < \infty\}) = 1.$

(iii) $\sum_n \int_{I_n} |c_n - y|\, m(dy) < \infty.$

Proof. We define

$$L_y = \sum_n \mathbf{1}_{I_n}(y)\, L^W(\mathrm{D}_{c_n}, y), \qquad y \in B.$$

By a result of Ray, Knight (cf. [35], 6.4) the law of the process $L^W(\mathrm{D}_{c_n}, \cdot)$ indexed by $y \in [x_0, c_n)$ if $c_n > x_0$ (resp. $y \in (c_n, x_0]$ if $c_n < x_0$) can be identified with the distribution of a squared 2-dimensional Bessel process R:

$$L^W(\mathrm{D}_{c_n}, y) \sim R^2_{|c_n - y|}.$$

Since a 2-dimensional Bessel process is just the radius of a 2-dimensional Wiener process, we have

$$L^W(\mathrm{D}_{c_n}, y)/|c_n - y| \sim \frac{1}{2}(X^2 + Y^2) \stackrel{\mathrm{def}}{=} \xi$$

where X and Y are independent standard Gaussian random variables. Hence, for $\phi(y) = \sum_n \mathbf{1}_{I_n}(y)|c_n - y|$, $y \in B$, it holds that

$$L_y/\phi(y) \sim \xi, \qquad y \in B.$$

Further, $I_n \subseteq [x_0, c_n)$ (resp. $I_n \subseteq (c_n, x_0]$) gives $\mathbf{P}(\{L_y = 0\}) = \mathbf{P}(\{\xi = 0\}) = 0$, $y \in B$.

Now, if for all n the condition

$$\begin{cases} \forall \varepsilon > 0 : \int_{I_n \cap [x_0, c_n - \varepsilon]} |c_n - y|\, m(\mathrm{d}y) < \infty & \text{if} \quad c_n > x_0 \\ \forall \varepsilon > 0 : \int_{I_n \cap [c_n + \varepsilon, x_0]} |c_n - y|\, m(\mathrm{d}y) < \infty & \text{if} \quad c_n < x_0 \end{cases}$$

is satisfied, then $\phi\, \mathrm{d}m$ is a σ-finite measure on $(B, \mathfrak{B}(B))$ and the assertion follows immediately from the preceding lemma.

Otherwise, if for some n and some $\varepsilon > 0$

$$\begin{cases} \int_{I_n \cap [x_0, c_n - \varepsilon]} |c_n - y|\, m(\mathrm{d}y) = \infty & \text{if} \quad c_n > x_0 \\ \int_{I_n \cap [c_n + \varepsilon, x_0]} |c_n - y|\, m(\mathrm{d}y) = \infty & \text{if} \quad c_n < x_0 \end{cases},$$

then we conclude as follows. Without loss of generality we consider only the case $c_n > x_0$. First, the above assumption yields $m(I_n \cap [x_0, c_n - \varepsilon]) = \infty$. From the RAY-KNIGHT representation of the local time as a squared Bessel process, we have $\mathbf{P}(\{L^W(\mathrm{D}_{c_n}, y) > 0\}) = 1$, $y \in [x_0, c_n - \varepsilon]$; moreover, $L^W(\mathrm{D}_{c_n}, \cdot)$ is $\mathbf{P}$-a.s. continuous. Thus, there exists a random variable K such that

$$L^W(\mathrm{D}_{c_n}, y) \geq K > 0, \qquad y \in [x_0, c_n - \varepsilon], \ \mathbf{P}\text{-a.s.}$$

This implies $\mathbf{P}$-a.s.

$$\begin{aligned} A &\geq \int_{I_n} L^W(\mathrm{D}_{c_n}, y)\, m(\mathrm{d}y) \\ &\geq \int_{I_n \cap [x_0, c_n - \varepsilon]} L^W(\mathrm{D}_{c_n}, y)\, m(\mathrm{d}y) \\ &\geq K\, m(I_n \cap [x_0, c_n - \varepsilon]) \\ &= \infty. \end{aligned}$$

Consequently, in this case it is evident that the assertions (i)-(iii) are equivalent. □

Now let $(W, \mathbb{G})$ on $(\Omega, \mathcal{F}, \mathbf{P}_x, x \in \mathbb{R})$ be a Wiener process (cf. (1.11)).

(A1.9) Lemma *Let m be a measure on $(K, \mathfrak{B}(K))$, where $K \in \mathfrak{B}(\mathbb{R})$ is fixed.*

(i) *If $m(K) > 0$ then $\int_K L^W(t, y)\, m(\mathrm{d}y) \to \infty$, $t \to \infty$, a.s.*

(ii) *If* $m(K) < \infty$ *then* $\int_K L^W(t,y)\,m(dy) < \infty$, $t \geq 0$, *a.s.*

Proof. (i) Using the well-known fact that

$$L^W(t,\cdot) \uparrow \infty \quad \text{if} \quad t \to \infty, \qquad \text{a.s.}$$

(cf. [35], ch. 3.6) the assertion is an easy consequence of the monotone convergence theorem.

(ii) $L^W(t,\cdot)$ vanishes outside the compact set $[\min_{s\leq t} W_s, \max_{s\leq t} W_s]$ (see (1.17)(i)) and, therefore, the a.s. continuity of $L^W(t,\cdot)$ implies that $L^W(t,\cdot)$ is bounded a.s. Hence the assertion follows immediately. □

(A1.10) Lemma *Let* (a,b) *be an interval and let* m *be a measure on* $((a,b), \mathfrak{B}((a,b)))$ *such that* $m(G) > 0$ *for all open subsets* $G \subseteq (a,b)$ *as well as* $m(K) < \infty$ *for all compact subsets* $K \subseteq (a,b)$. *We set*

$$B_t = \int_{(a,b)} L^W(t,y)\,m(dy), \qquad t \geq 0,$$

and fix a finite $\mathbb{G}$*-stopping time* η. *Then* $B_.(\omega)$ *is strictly increasing on* $[\eta(\omega), \eta(\omega) + D_{(a,b)^c} \circ \theta_{\eta(\omega)}(\omega))$ *for* $\omega \in \{W_\eta \in (a,b),\, B_\eta < \infty\}$ *a.s.*[2]

Proof. Denote $A = \{(s,\omega) | \exists \varepsilon > 0 : B_s(\omega) = B_{s+\varepsilon}(\omega)\}$ and fix $x \in \mathbb{R}$. Obviously, the set A is $\mathbb{F}^W$–optional. We have to show that $A \cap [\![\eta, \eta + D_{(a,b)^c} \circ \theta_\eta[\![\cap ([0,\infty) \times \{W_\eta \in (a,b), B_\eta < \infty\})$ is $\mathbf{P}_x$-evanescent (cf. [12]). By the optional section theorem (cf. [12], IV T 10), for this it suffices to prove that for every stopping time ξ with $[\![\xi]\!] \subseteq [\![\eta, \eta + D_{(a,b)^c} \circ \theta_\eta[\![\cap ([0,\infty) \times \{W_\eta \in (a,b), B_\eta < \infty\})$ it holds that

$$B_{\xi+\varepsilon} > B_\xi,\ \varepsilon > 0$$

$\mathbf{P}$-a.s. on $\{\xi < \infty\}$.

Now applying the strong Markov property,

$$\mathbf{P}_x(\{B_{\xi+\varepsilon} > B_\xi,\ \forall\varepsilon > 0\} \cap \{\xi < \infty\}) = \mathbf{E}_x \mathbf{1}_{\{\xi<\infty\}} \mathbf{P}_{W_\xi}(\{B_\varepsilon > 0,\ \forall\varepsilon > 0\}).$$

Thus it suffices to prove that $\mathbf{P}_y(\{B_\varepsilon > 0,\ \forall\varepsilon > 0\}) = 1$, $y \in (a,b)$. However, as is well-known and also an easy consequence of (1.17)(vi), we have that $\mathbf{P}_y(\{L^W(\varepsilon,y) > 0, \varepsilon > 0\}) = 1$. Now the continuity of $L^W(\varepsilon,\cdot)$ yields that there is a neighbourhood U of y and a $K > 0$ (both depending on ω) such that

$$L^W(\varepsilon,z) \geq K > 0, \qquad z \in U.$$

Because of $m(U) > 0$ this proves the assertion. □

[2] Without loss of generality we suppose the existence of shift operators for W.

Appendix 2

Nonnegative Functions of Finite Variation and Associated Signed Measures

Let R be an open subset of the real axis $\mathbb{R}$ and let $g : R \to [0, \infty)$ be a right-continuous function which is of bounded variation on each compact subinterval of R. We define $\bar{g}(x) = \frac{1}{2}(g(x) + g(x-))$, $x \in R$.

In this Appendix we will elaborate the formal expression "$(\bar{g})^{-1} dg$". In general, because of the zeros of $\bar{g}$, this expression cannot be interpreted as a signed measure on $(R, \mathcal{B}(R))$. For example, if $g(x) = |x|^{\frac{1}{2}}$, $x \in \mathbb{R}$, then it holds that

$$\int_{-\varepsilon}^{0} (\bar{g})^{-1} dg = -\infty, \quad \int_{0}^{\varepsilon} (\bar{g})^{-1} dg = \infty$$

for all $\varepsilon > 0$.

Let $R = \bigcup_n I_n$ denote the decomposition of R into its components. It is well-known that g can be represented as

$$g = g_1 - g_2,$$

where the functions g_i are increasing and right-continuous on I_n. Moreover, there exist disjoint sets $A_1, A_2 \in \mathcal{B}(R)$ such that $A_1 \cup A_2 = R$ and

$$\int_{A_1} dg_2 = \int_{A_2} dg_1 = 0.$$

In what follows, concerning the topology on R we refer to the trace topology induced by the ordinary topology on $\mathbb{R}$. Denote

$$M_g = \left\{ x \in R : \int_U (\bar{g})^{-1} dg_1 = \infty \text{ or } \int_U (\bar{g})^{-1} dg_2 = \infty \right.$$
$$\left. \text{for all neighbourhoods } U \text{ of } x \right\}.$$

(A2.1) Proposition (i) *M_g is closed in R.*

(ii) $M_g = \{x \in R \,|\, g(x) = 0 \textit{ and } \forall \varepsilon > 0 \, \exists y \in (x, x+\varepsilon) \cap R : \bar{g}(y) > 0\} \cup \{x \in R \,|\, g(x-) = 0 \textit{ and } \forall \varepsilon > 0 \, \exists y \in (x-\varepsilon, x) \cap R : \bar{g}(y) > 0\}.$

Proof. (i) This property is an immediate consequence of the definition of M_g.

(ii) Let $x_0 \in M_g$ as well as $\int_U (\bar{g})^{-1} \mathrm{d}g_1 = \infty$ for all neighbourhoods U of x_0. If $g(x_0) > 0$ and $g(x_0-) > 0$ then there is a sufficiently small neighbourhood U of x_0 such that $g(y) \geq c > 0$, $g(y-) \geq c > 0$, $y \in U$, which contradicts $x_0 \in M_g$. Otherwise, if $g(x_0) = 0$ (resp. $g(x_0-) = 0$) and if there exists an $\varepsilon > 0$ such that $\bar{g}$ vanishes on $(x_0, x_0+\varepsilon) \cap R$ (resp. $(x_0-\varepsilon, x_0) \cap R$) then, $g_1 = g_2 =$ constant on $[x_0, x_0+\varepsilon) \cap R$ (resp. $(x_0-\varepsilon, x_0) \cap R$) follows. This implies

$$\int_{[x_0, x_0+\varepsilon)\cap R} (\bar{g})^{-1} \mathrm{d}g_i = 0$$

$$\left(\text{resp. } \int_{(x_0-\varepsilon, x_0)\cap R} (\bar{g})^{-1} \mathrm{d}g_i = 0, \int_{(x_0-\varepsilon, x_0]\cap R} (\bar{g})^{-1} \mathrm{d}g_i < \infty \right),$$

$i = 1, 2$, leading to a contradiction of $x_0 \in M_g$, again.

Now, assume that x_0 is an element of the right-hand side of (ii). By symmetry, we only deal with the case that $g(x_0) = 0$ and $\forall \varepsilon > 0 \, \exists y \in (x_0, x_0+\varepsilon) \cap R : \bar{g}(y) > 0$. Fix $\varepsilon > 0$ such that $(x_0, x_0+\varepsilon) \subseteq R$. Without loss of generality we may suppose that $g_1(x_0) = g_2(x_0) = 0$. Then $\bar{g} \leq \bar{g}_1$ on $(x_0, x_0+\epsilon)$ and to show that $x_0 \in M_g$ it suffices to prove that

$$\int_{(x_0, x_0+\varepsilon)\cap R} (\bar{g}_1)^{-1} \mathrm{d}g_1 = \infty.$$

Therefore, in what follows we will assume that g is increasing on $[x_0, x_0+\varepsilon)$, $g(x_0) = 0$ as well as $g > 0$ on $(x_0, x_0+\varepsilon)$.

Using the right-inverse h of g on $[x_0, x_0+\varepsilon)$ defined by

$$h(y) = \inf\{x \geq x_0 : g(x) > y\}, \qquad y \in [0, g(x_0+\varepsilon)),$$

we obtain

$$\int_{(x_0, x_0+\varepsilon)} (\bar{g})^{-1} \mathrm{d}g \geq \int_{(x_0, x_0+\varepsilon)} g^{-1} \mathrm{d}g = \int_0^{g(x_0+\varepsilon)} (g \circ h(x))^{-1} \mathrm{d}x.$$

Let us investigate the function $g \circ h$ on the interval $J = [0, g(x_0+\varepsilon))$. Clearly, the set $B = \{x \in J : h(x+\delta) > h(x) \text{ for all } \delta > 0\}$ is closed to the left. Hence, $J \setminus B$ is at most a countable union of disjoint intervals $J_n = [c_n, d_n)$, $c_n < d_n$. On the set B it holds that $g \circ h(x) = x$ and for $x \in J_n$ we compute $g \circ h(x) = d_n$. Therefore, we obtain

$$\begin{aligned} \int_{(x_0, x_0+\varepsilon)} (\bar{g})^{-1} \mathrm{d}g &\geq \int_B x^{-1} \, \mathrm{d}x + \sum_n \int_{J_n} d_n^{-1} \, \mathrm{d}x \\ &\geq \int_B x^{-1} \, \mathrm{d}x + \sum_n \left(1 - \frac{c_n}{d_n}\right). \end{aligned}$$

We finish the proof of (ii) by showing that at least one of the above summands diverges. If $\sum_n (1 - \frac{c_n}{d_n})$ only consists of a finite number of terms, say n_0, then the following inequality is evident:

(A2.2)
$$\int_{(x_0,x_0+\varepsilon)} (\bar{g})^{-1} \mathrm{d}g \geq \frac{1}{2} \int_{J \setminus (J_1 \cup \ldots \cup J_{n_0})} x^{-1} \, \mathrm{d}x.$$

On the other hand, if the series possesses infinitely many terms and converges, then $\frac{c_n}{d_n} \to 1$, $n \to \infty$ holds and because of $\frac{c_n}{d_n} < 1$ we have $\frac{c_n}{d_n} \in (\frac{1}{2}, 1)$ for all $n \geq n_0$, n_0 sufficiently large. Using $2(1-x) \geq -\ln(x)$, $x \in (\frac{1}{2}, 1)$, we derive

$$\sum_{n>n_0} \left(1 - \frac{c_n}{d_n}\right) \geq \frac{1}{2} \sum_{n>n_0} \left(-\ln \frac{c_n}{d_n}\right) = \frac{1}{2} \sum_{n>n_0} \int_{c_n}^{d_n} x^{-1} \, \mathrm{d}x$$

leading to (A2.2), again.

Now, we claim that $J \setminus (J_1 \cup \ldots \cup J_{n_0})$ includes an interval $(0, \delta)$ for some $\delta > 0$. We verify this by showing that $c_n \neq 0$, $n = 1, \ldots, n_0$. Indeed, if $c_n = 0$ then h is constant on $[0, d_n)$: $h(x) = c$, $x \in [0, d_n)$. Now if $c = x_0$ then $g(x_0+) > 0$ which is a contradiction of $g(x_0+) = 0$. Otherwise, if $c > x_0$ then g must be constant in a right neighbourhood of x_0 contradicting to $g > 0$ on $(x_0, x_0 + \varepsilon)$. Summarizing, we obtain

$$\int_{(x_0,x_0+\varepsilon)} (\bar{g})^{-1} \mathrm{d}g \geq \frac{1}{2} \int_{(0,\delta)} x^{-1} \, \mathrm{d}x = \infty. \qquad \square$$

Now, on certain subsets of R we will establish a correspondence between the formal expression

$$(\bar{g})^{-1} \mathrm{d}g$$

and a signed measure. Here it is clear that the points of M_g are in some sense critical. Because of (A2.1)(i) the set $R \setminus M_g$ is open and, hence, it can be decomposed into components $R_n^0 = (a_n, b_n)$, $a_n < b_n$. We set

$$R_n = R_n^0 \cup (\{a_n\} \cap \{g > 0\}) \cup (\{b_n\} \cap \{g_- > 0\})^1$$

and

$$R_g = \bigcup R_n.$$

(A2.3) Lemma (i) *For $m \neq n$ the sets R_m and R_n are disjoint: $R_n \cap R_m = \emptyset$.*

(ii) *$R \setminus R_g$ is included in $\{\bar{g} = 0\}$ and for each compact subset K of R it holds that*

$$0 = \int_{\{\bar{g}=0\} \cap K} |\mathrm{d}g| = \int_{(R \setminus R_g) \cap K} |\mathrm{d}g|.$$

---

[1] $g_-(x) = g(x-)$.

Proof. (i) immediately follows from (A2.1)(ii).

(ii) Let $x \in R \setminus R_g$. At first, this implies $x \in M_g$. Now if $g(x-) > 0$ then there exists a $\delta > 0$ and a constant $c > 0$ such that $g(y) > c$, $g(y-) > c$ for all $y \in (x-\delta, x) \cap R$ leading to $(x-\delta, x) \subseteq R \setminus M_g$ by (A2.1)(ii). But, because of $x \in M_g$, this yields $x = b_n$ for some n and, consequently, since $g(x-) > 0$, this implies $x \in R_n \subseteq R_g$ which is a contradiction. Analogously we deal with the case $g(x) > 0$.

It remains to prove that for every component I_i of R and for every $N > 0$

$$\int_{\{\bar{g}=0\}\cap[-N,N]\cap I_i} |\mathrm{d}g| = 0.$$

To simplify the notation we assume that $R = I_i = \mathbb{R}$. Let $g = g^c + g^d$ be the decomposition of g into the sum of a continuous function g^c and a pure jump function g^d. In view of $g \geq 0$ the set $\{\bar{g} = 0\}$ contains no point of discontinuity of g, thus

$$\int_{\{\bar{g}=0\}\cap[-N,N]} |\mathrm{d}g| = \int_{\{\bar{g}=0\}\cap[-N,N]} |\mathrm{d}g^c|.$$

Since g is locally of bounded variation, it admits at most a countable number of discontinuities and g^d takes at most a countable number of values $a_1, \dots, a_n, \dots$ For $x \in \{\bar{g} = 0\}$ we have $-g^c(x) = g^d(x+) = g^d(x-)$ and therefore $\{\bar{g} = 0\} \subseteq \bigcup_n \{g^c = -a_n\}$. Consequently it suffices to verify that

$$\int_{\{g^c=-a_n\}\cap[-N,N]} |\mathrm{d}g^c| = 0$$

for every n. Let n be fixed and let A be a closed subset of $\{g^c = -a_n\} \cap [-N, N]$. The complement A^c of A is the union of disjoint intervals:

$$A^c = \bigcup_k (c_k, d_k) \text{ with } g^c(c_k) = g^c(d_k) = -a_n \text{ if } c_k > -\infty,\ d_k < \infty.$$

Then we have

$$\begin{aligned} \int_A \mathrm{d}g^c &= \int_{[-N,N]} \mathrm{d}g^c - \int_{[-N,N]\cap A^c} \mathrm{d}g^c \\ \textbf{(A2.4)} \qquad &= g^c(N) - g^c(-N) - \sum_k \int_{[-N,N]\cap(c_k,d_k)} \mathrm{d}g^c \\ &= 0. \end{aligned}$$

Now let $\mathrm{d}g^c = \mu_1 - \mu_2$ denote the Jordan decomposition of the signed measure generated by g^c on $([-N,N], \mathcal{B}([-N,N]))$, i.e. μ_1, μ_2 are finite measures on $([-N,N], \mathcal{B}([-N,N]))$ and there exists a set $D \in \mathcal{B}([-N,N])$ such that $\mu_1(D) = \mu_2([-N,N] \setminus D) = 0$. Equality (A2.4) now yields $\mu_1(A) = \mu_2(A)$ for every closed subset A of $\{g^c = -a_n\} \cap [-N,N]$ which in turn implies

$\mu_1(B \cap \{g^c = -a_n\}) = \mu_2(B \cap \{g^c = -a_n\})$ for every $B \in \mathcal{B}([-N, N])$. Applying this for $B = D$ and $B = D^c$ we obtain $\mu_1([-N, N] \cap \{g^c = -a_n\}) = \mu_2([-N, N] \cap \{g^c = -a_n\}) = 0$ and thus

$$\int_{\{g^c=-a_n\}\cap[-N,N]} |dg^c| = 0.$$

□

(A2.5) Lemma *If $K \subseteq R_n$ is a compact subinterval then*

$$\mu_g(A) = \int_A (\bar{g})^{-1} dg, \qquad A \in \mathfrak{B}(K),$$

defines a finite signed measure on $(K, \mathfrak{B}(K))$.

Proof. Let $K = [c, d]$. Then (A2.1) and the definition of R_n imply $g(c) > 0$ (resp. $g(d-) > 0$) or $\bar{g} = 0$ on $(c, c + \varepsilon)$ (resp. $(d - \varepsilon, d)$) for some $\varepsilon > 0$. All in all, this yields

$$\int_{[c,c+\varepsilon)} (\bar{g})^{-1} dg_i < \infty$$

and

$$\int_{(d-\varepsilon,d]} (\bar{g})^{-1} dg_i < \infty,$$

$i = 1, 2$, ε sufficiently small. Using the definition of R_n, for all points x of the remaining interval $[c + \varepsilon, d - \varepsilon]$ there exists a neighbourhood U_x such that

$$\int_{U_x} (\bar{g})^{-1} dg_i < \infty, \qquad i = 1, 2.$$

But, the compact interval $[c + \varepsilon, d - \varepsilon]$ can already be covered by a finite number of neighbourhoods U_x proving the assertion. □

Now, we approximate each R_n by compact subintervals R_n^k, $k = 1, 2, \ldots$: $R_n^k \uparrow R_n$, $k \to \infty$. Clearly, if R_n is itself compact we only need to set $R_n^k = R_n$ for all k, sufficiently large. Introduce the set function

(A2.6) $$\mu_g(A) = \int_A (\bar{g})^{-1} dg, \; A \in \bigcup_{n,k} \mathfrak{B}(R_n^k),$$

which is well-defined by (A2.5). Finally, taking into account (A2.3) we are able to give sense to the formal expression "$(\bar{g})^{-1} dg$".

(A2.7) Definition A measurable function $f : R \to \mathbb{R}$ is said to be μ_g-*integrable* if

$$\sum_n \lim_k \int_{R_n^k} |f|\,|\mathrm{d}\mu_g| < \infty.$$

Then we define

$$\int_R f\,\mathrm{d}\mu_g = \sum_n \left[\lim_k \int_{R_n^k} f(\bar{g})^{-1}\mathrm{d}g_1 - \lim_k \int_{R_n^k} f(\bar{g})^{-1}\mathrm{d}g_2\right].$$

Obviously, this definition of $\int_R f\,\mathrm{d}\mu_g$ does not depend on a particular choice of the approximating sequence $(R_n^k)_{k\in\mathbb{N}}$ of compact subintervals.

(A2.8) Proposition *A measurable function $f : R \to \mathbb{R}$ is μ_g-integrable if and only if*

$$\int_R |f|(\bar{g})^{-1}\,|\mathrm{d}g| < \infty^2.$$

Then we have

$$\int_R f\,\mathrm{d}\mu_g = \int_R f(\bar{g})^{-1}\mathrm{d}g.$$

The proof of this assertion immediately follows from (A2.7) and (A2.3)(ii).

---

[2] Remind (1.28).

References

1. S. Assing, On Reflected Solutions of Stochastic Differential Equations with Ordinary Drift, *Stochastics and Stochastics Reports* **42**, 183-198 (1993)
2. S. Assing, *Homogene Differentialgleichungen mit Gewöhnlicher Drift*, Thesis, Friedrich-Schiller-Universität Jena, 1994
3. J. Azema, M. Yor (Eds.), Temps locaux, *Asterisque* **52-53**, 1978
4. K. Bichteler, Stochastic Integrators, *Bull. American Math. Soc.* **1**, 761-765 (1979)
5. K. Bichteler, Stochastic Integration and L^p-theory of semimartingales, *Annals of Probability* **9**, 49-89 (1981)
6. R.M. Blumenthal, R.K. Getoor, *Markov Processes and Potential Theory*, Academic Press, New York, London, 1968
7. M. Chaleyat-Maurel, N. El Karoui, Un problème de réflexion et ses applications au temps local et aux équations différentielles stochastiques sur $\mathbb{R}$ - Cas continu, *Astérisque*, Société Mathématique de France, **52-53**, 117-144 (1978)
8. K.L. Chung, R.J. Williams, *Introducion to Stochastic Integration*, Birkhäuser, Boston 1990
9. E. Cinlar, Markov additive Processes I, II, *Z. Wahrscheinlichkeitstheorie verw. Geb.* **24**, 85-93, 94-121 (1972)
10. E. Cinlar, J. Jacod, Representation of Semimartingale Markov Processes in Terms of Wiener Processes and Poisson Random Measures, In: Seminar on Stochastic Processes, *Progress in Probability and Statistics Vol.* **1**, 159-242, Birkhäuser 1981
11. E. Cinlar, J. Jacod, P. Protter, M.J. Sharpe, Semimartingales and Markov Processes, *Z. Wahrscheinlichkeitstheorie verw. Geb.* **54**, 161-219 (1980)
12. C. Dellacherie, *Capacités et processus stochastiques*, Springer 1972
13. C. Dellacherie, Un survol de la théorie de l'intégrale stochastique, *Stochastic Processes and Their Applications* **10**, 115-144 (1980)
14. C. Dellacherie, P.A. Meyer, *Probabilités et Potentiel*, Chaps. V-VIII, Herrmann, Paris 1980
15. J. L. Doob, Martingales and one-dimensional diffusion, *Trans. of the AMS* **78**, 168-208 (1955)
16. E. B. Dynkin, *Markov processes*, Springer 1965
17. N. El Karoui, Sur les montées des semi-martingales, Temps locaux, *Asterisque* **52-53**, 63-87 (1978)
18. H.J. Engelbert, Markov Processes in General State Spaces (Part I) *Math. Nachr.* **80**, 19-36 (1977); (Part II) *Math. Nachr.* **82**, 191-203 (1978); (Part IV) *Math. Nachr.* **84**, 277-300 (1978)
19. H.J. Engelbert, On the Strong Markov Property of One-Dimensional Continuous Markov Processes, In: Stochastic Processes and Optimal Control, Proceedings of the 9th Jena Winterschool on Stochastic Processes held in Friedrichroda 1992, *Stochastic Monographs* **7**, 51-71 (1993)

20. H.J. ENGELBERT, W. SCHMIDT, On one-dimensional stochastic differential equations with generalized drift, In: Stochastic Differential Systems, Proceedings, *Lecture Notes in Control and Information sciences* **69**, 143-155 (1985)
21. H.J. ENGELBERT, W. SCHMIDT, Strong Markov continuous local martingales and solutions of one-dimensional stochastic differential equations (Part I), *Math. Nachr.* **143**, 167-184 (1989)
22. H.J. ENGELBERT, W. SCHMIDT, Strong Markov continuous local martingales and solutions of one-dimensional stochastic differential equations (Part II), *Math. Nachr.* **144**, 241-281 (1989)
23. H.J. ENGELBERT, W. SCHMIDT, Strong Markov continuous local martingales and solutions of one-dimensional stochastic differential equations (Part III), *Math. Nachr.* **151**, 149-197 (1991)
24. H.J. ENGELBERT, W. SCHMIDT, On Solutions of One-Dimensional Stochastic Differential Equations Without Drift, *Z. Wahrscheinlichkeitstheorie verw. Geb.*, **68**, 287-314 (1985)
25. W. FELLER, The General Diffusion Operator and Positivity Preserving Semigroups in One Dimension, *Ann. Math.* **60**, 417-436 (1954)
26. W. FELLER, On Second Order Differential Operators, *Ann. Math.* **61**, 90-105 (1955)
27. W. FELLER, Generalized Second Order Differential Operators and their Lateral Conditions, *Illinois J. Math.* **1**, 495-504 (1957)
28. D. FREEDMAN, *Brownian motion and diffusion*, Holden-Day, San Francisco, 1971
29. H. IKEDA, S. WATANABE, *Stochastic Differential Equations and Diffusion Processes*, North-Holland, 1981
30. K. ITÔ, On Stochastic Differential Equations, *Mem. Amer. Math. Soc.* **4**, 1-51 (1951)
31. K. ITÔ, H. P. MCKEAN, *Diffusion processes and their sample paths*, Springer 1965
32. J. JACOD, *Calcul stochastique et problèmes de martingales*, Lecture Notes in Mathematics **714**, 1979
33. T. JEULIN, *Semi-martingales et grossissement d'une filtration*, Lecture Notes in Mathematics **833**, 1980
34. S. KARLIN, H.M. TAYLOR, *A Second Course in Stochastic Processes*, Academic Press 1981
35. I. KARATZAS, S.E. SHREVE, *Brownian Motion and Stochastic Calculus*, Springer 1987
36. H. KUNITA, S. WATANABE, On square-integrable martingales, *Nagoya Math. J.* **30**, 209-245 (1967)
37. J.F. LEGALL, *Temps locaux et équations différentielles stochastiques*, Thèse 3ème cycle, Université de Paris VI, 1982
38. S. MÉLÉARD, Application du calcul stochastique à l'étude de processus de Markov réguliers sur [0,1], *Stochastics* **19**, 41-82 (1986)
39. P. A. MEYER, Un cours sur les intégrales stochastiques, Séminaire Probab. X, *Lecture Notes in Mathematics* **511**, 246-400 (1976)
40. P. PROTTER, *Stochastic Integration and Differential Equations*, Springer 1992
41. L.C.G. ROGERS, D. WILLIAMS, *Diffusions, Markov Processes, and Martingales, Vol.* **2**: *Itô Calculus*, Wiley, New York 1987
42. W. SCHMIDT, On stochastic differential equations with reflecting barriers, *Math. Nachr.* **142**, 135-148 (1989)
43. W. SCHMIDT, On semimartingale diffusions and stochastic differential equations, *Stochastics and Stochastics Reports* **29**, 407-424 (1990)

44. W. SCHMIDT, Weakly additive functionals and time change of strong Markov processes, In: Stochastic Processes and related Topics, *Mathematical Research* **61**, 145-152, Akademie-Verlag, Berlin, 1991
45. C. Stricker, M. Yor, Calcul stochastique dépendant d'un paramètre, *Z. Wahrscheinlichkeitstheorie verw. Geb.* **45**, 109-133 (1978)
46. V. A. VOLKONSKI, Random substitution of time in strong Markov processes, *Teor. Veroyatnost. i ee Primenen.* **3**, 332-350 (1958)
47. J. B. WALSH, The Perfection of Multiplicative Functionals, In: Séminaire de Probabilités VI, *Lecture Notes in Mathematics* **258**, 233-242 (1972)
48. A. T. WANG, Generalized Itô's formula and additive functionals of Brownian motion, *Z. Wahrscheinlichkeitstheorie verw. Geb.* **41**, 153-159 (1977)
49. D. WILLIAMS, *Diffusions, Markov Processes, and Martingales. Vol.* **1**: *Foundations*, J. Wiley & Sons, Chichester 1979
50. X. -X. XUE, A Zero-One Law for Integral Functionals of the Bessel Process, In: Séminaire de Probabilités XXIV, *Lecture Notes in Mathematics* **1426**, 137-153 (1990)
51. M. YOR, Sur la continuité des temps locaux associés à certaines semimartingales, In: Temps locaux, *Asterisque* **52-53**, 23-35 (1978)

Index

Symbols